GAME TESTING

LICENSE, DISCLAIMER OF LIABILITY, AND LIMITED WARRANTY

Companion files for this title are available with proof of purchase by writing to the publisher at info@merclearning.com.

Game Testing

All In One

Fourth Edition

Robert Denton Bryant

Mercury Learning and Information
Boston, Massachusetts

Publisher: David Pallai
MERCURY LEARNING AND INFORMATION
121 High Street, 3rd Floor
Boston, MA 02110
info@merclearning.com
www.merclearning.com
800-232-0223

Robert Denton Bryant. *Game Testing: All in One, Fourth Edition.*
ISBN: 978-1-50152-168-3

Library of Congress Control Number: 2024932259

242526321 This book is printed on acid-free paper in the United States of America.

Our titles are available for adoption, license, or bulk purchase by institutions, corporations, etc. For additional information, please contact the Customer Service Dept. at 800-232-0223(toll free).

All of our titles are available in digital format at academiccourseware.com and other digital vendors. *Companion files for this title are available with proof of purchase by contacting info@merclearning.com.* The sole obligation of MERCURY LEARNING AND INFORMATION to the purchaser is to replace the book/files, based on defective materials or faulty workmanship, but not based on the operation or functionality of the product.

Contents

FOREWORD

Having known Bob Bryant for many years, I am honored to be writing the foreword for this fourth edition of *Game Testing: All in One*. That the first edition was in 2005 and almost two decades later we are into the fourth edition is remarkable.

With the challenges that Game QA faces today, to put it bluntly, we are long past the point where simply adding more people to our test teams for coverage is the answer.

Then why do we need this book if simply adding more people is not the answer? We still need people, many more people. It is just that the requirements of the role are changing.

Game QA is still one of the best proving grounds for growing the future leaders of tomorrow's game studios that I know. More so now than ever, it is a stand-alone discipline within game production. Yet many studios continue to underestimate its criticality to not just development, publishing, and launch, but also to the key metric of player retention.

The discipline of Game QA is becoming ever more complex and the ways in which we work are increasingly more sophisticated. It is now its own career path within the game production ecosystem.

Many game studios have dropped the term "Quality Assurance" and now simply use the word "Quality." A recent sample of testing job roles across the industry included such job titles as quality design, quality verification, quality control, test analysis, quality data analysis, quality engineering, technical quality, compliance, compatibility, release verification, network testing, and user acceptance. The list of terms goes on and on.

The current challenges we face as Game QA are manifold, but can be broken down into two clear areas. The first of these is the volume of content and code that is now generated at an increasingly accelerated pace and the sheer complexity of the game titles that are released.

This can be addressed with automation and tools. Automation will ensure we have coverage over any previously released content and that new content has not broken the "last known good," to paraphrase a Microsoft term.

Tools help us to work faster and more efficiently. The term artificial intelligence (AI) was coined in the 1950s and is based on the idea of Intelligence Amplification. I believe our current tools future is Augmented AI.

These are the tools we can use to reinforce our decision-making, validate our thinking, and which will enable us to work smarter and more effectively, but not replace us.

Automation ensures that we have not broken what was previously released. However, new content still requires testing by our test teams. More so now than ever, "finding the fun" is a critical part of the Quality role.

Finding exploits is critical: escaped defects and exploits are the costliest bugs a studio can let slip through, with not just revenue implications but also reputational damage and impact to the latest production schedules.

The concept of "Shift Left," although not new, is also gaining prominence. Quality is becoming about prevention and not detection. Quality becomes an integral part of the design process, quality built in throughout production and everyone being responsible for quality across the studio.

The second challenge is the very nature of "live service" games themselves and the size of the player communities we now serve. Games are no longer shipped with a "fire and forget" attitude. They are products that last years, decades even, building social networks within their communities for our players.

Players put in more play time in the first week, days, hours, or even minutes after launch than the time that we can spend testing the games internally, even with hundreds of testers working for months before launch!

At this juncture, we should consider two questions. What is the role of Game QA? How has this changed?

QA is often referred to as the voice of the player. I believe that in the "games as a service" world, our test teams are the bridge to the player, where at all points in the game production process we are building in quality feedback loops across internal teams and our player communities.

Does the QA role go beyond prevention? Is it also about listening to what our players want, what brings them joy, and what detracts from their experience?

I do not have all the answers to these questions. The customer is usually right, as the adage goes, but reacting to those players who shout the loudest is not the best approach either!

My conclusion, though, is that it is probably more about listening to the voice of everyone in our communities, analyzing what that data is showing us, and making informed quality decisions.

The roles and responsibilities of QA have developed exponentially over the past two decades, the complexities of the games we deal with require many more roles and responsibilities. The basics of game testing, however, have not.

This to me is why this book remains crucial for both entry-level testers, test managers, developers or anyone involved in production, who wants to understand the principles of one of the critical disciplines required for successfully shipping a high-quality game.

Ben Wibberley
Cambridge, England

Ben Wibberley is VP of Game Operations at Jagex and has over twenty-five years of experience in the video game industry. Having started his career as a game tester, he has dedicated his career to championing Game QA. He co-founded Babel Media, the first multi-service game outsourcing provider, in 1998. He also set up a publishing support network at DDM and ran global business development for VMC. He founded DAQA, the Concierge of Game Quality, a boutique game quality studio. He is a co-founder of GameQuality.org and a co-chair of the Qualicon content committee. He will always consider himself to be a tester, albeit not a very good one.

Preface to the Fourth Edition

The first edition of this book appeared almost twenty years ago. (Spacetime is relentless like that.) Since then, the world of video games has both transformed and exploded. More players are playing games than ever before, and there are more platforms for playing on and more ways of making money from games than ever before. Small wonder, then, that some game testers have started to organize and join labor unions in the last few years. This work is skilled labor, and game testers deserve to be adequately compensated for their hard work, enjoying the same benefits and job protections as other members of the game development team.

As we drafted the first edition of *Game Testing All in One*, Charles P. Schultz and I described the best practices of game testing at a time when most video games in North America and Europe were sold at retail stores and played on computers or game consoles, be they handheld or attached to a television. Now most games are downloaded to smartphones, smart TVs, consoles, and computers. Billion-dollar global companies release years-in-the-making "AAA" games that are epic in their scope. Tiny one-person development teams release quirky little mobile games. Billions of dollars have been spent in the last several years by thousands of companies trying to make a market for virtual reality. *Pokémon Go* proved the market for augmented reality games in a weekend. The global pandemic that began in 2020 supercharged the growth of video games, as players worldwide turned to titles like *Animal Crossing* to escape the grubby worries of actual reality.

There have never been more games competing for the attention of more players. Never before has game quality and stability been more important. The process and discipline of game testing is crucial for any development team of any size to learn and adhere to, especially in this new era of constant patches, updates, feature roll-outs, expansions, and DLC releases. It seems today that for so many games, development is never finished. Neither, then, is testing.

Fortunately, the fundamental skills involved in game testing have changed very little. What has changed in many cases is the complexity of the games tested, the number of platforms they are required to be tested on, and the number of ways players can interact with—and spend money in—a game. Although automation and artificial intelligence can help to make the game tester's job easier, gameplay has to feel right to human players, so human testers will always be needed.

For this fourth edition, we have revised and restructured the book to make it more useful to students, teachers, entry-level testers, and those new to video game development. We have restored and updated material describing the overall game development process and the essential role that quality assurance (QA) plays within it. We have moved to the second half of the book those chapters aimed at test managers and game producers who want to improve the efficiency and efficacy of their testing resources. And we have added a glossary!

Software engineers will continue to make mistakes. Game designers will inadvertently introduce exploits to their carefully crafted gameplay systems. Artists in haste may fail to close vertices or hide seams. It remains the job of the game tester to advocate for the players by "breaking" the game before the player ever gets to play it. It is up to vigilant game testers to save the player from frustration, confusion, and lost time, and thereby help to ensure the commercial and critical success of the game. Like the Night's Watch in *Game of Thrones*, testers are the anonymous, out-of-sight, and unsung heroes of the realm of video game development.

Our hope is that this upgraded edition of the book will help you to develop your skills as a tester, or test manager, so that you can remain vigilant. The players, though they may not know it, are ever grateful for your hard work.

Most sadly, this is the first edition of the book to be published without new material from my brother in testing, Charles P. Schultz, who passed away in November of 2022.

This book would not exist without his inspiration, vision, and leadership. Rest in peace, my friend.

Robert Denton Bryant
Kyle, TX
April 2024

ACKNOWLEDGMENTS

It would be a "show-stopping bug" if I failed to acknowledge the valuable support and input of my colleagues and students at St. Edward's University in Austin, Texas, most particularly, Assistant Professor of Video Game Development, Jeremy Johnson, PhD.

I am also very grateful to David Pallai at Mercury Learning and Information, Heather Maxwell Chandler, Stefan Seicarescu, and Marius Popa at Quantic Lab, and Terri DePaolo, whose patience is boundless.

Very special thanks to Mirta Schultz, for helping to make this edition of the book possible.

CHAPTER 1

Your Role on the Game Development Team

"YOU GET PAID TO PLAY GAMES?"

The job of testing video games is very hard, very rewarding, and very weird. When you tell friends what you do, they may ask, enviously, "You get paid to play games?"

"No," you will reply. "Game testing is hard work. It's a real job."

The first time you are assigned to test a game, you might play it as a player would, feeling the joy of discovery as you learn the controls, setting, and gameplay loop.

But the fortieth time you launch the same game, it is *work*.

The discipline of game testing, often referred to as *quality assurance* or QA, is a crucial aspect of the process of video game development. Game testing is (and should always be) considered an integral part of the wide range of skills needed to produce a successful video game.

Game teams come in different sizes, structures, and locations. They can vary by company, game genre, or platform. The different disciplines required to produce a working video game are often organized into distinct departments within the overall game team. The people on the team must work together to complete the tasks that are needed to get the game done on time and without serious software defects, or *bugs*. By understanding the different departments on a typical video game project—even though developers may insist that their project is anything but typical—you can begin to visualize your role within the QA department and how it contributes to the success of the project.

ELEMENTS OF A GAME DEVELOPMENT TEAM

Programmers

Programmers make the video game work. They use their programming skills to turn the game design into something you can play on your computer or other device. When unexpected problems arise, they are quick to fix them with their problem-solving skills. They also know the details of the software tools used to make the game, which helps them to improve performance and fix tricky bugs.

The members of this team may call themselves programmers, engineers, coders, scripters, or developers. (This last term is falling out of use, however, since anyone who works for a game development company can be considered a game developer.) They produce the code that the testers must test before the game can be released. Their job involves translating the gameplay, artwork, sound, and story of the game into programming language. This code is subsequently converted into a game program that players run on their devices.

The code has to fit within a certain budget, use a limited amount of working memory, and be able to make things such as user input responses and video frame rate flow smoothly. The game program also has to fit within whatever development framework the team is using. This may include elements such as a game *engine*, which follows defined rules for automating the processing of certain game elements, and *middleware*, which provides a common interface to certain game functions so the same code can be moved from one platform to another. Programmers also have to deal with operating systems, device drivers, and communications protocols for multiplayer games, each with their own complexities and challenges.

Programmers may be involved in the porting of game code from one platform to another. Certain portions of the code should remain the same, whereas others must be changed to accommodate differences in the new platform. Porting a game from Windows PC to Xbox may not be difficult because those platforms are similar. However, the PC version needs to account for a variety of screen resolutions, graphics cards, installed memory, audio devices, and input devices. Going from Xbox to another game console would provide a different set of challenges, and porting a game from a game console to a tablet or mobile device may result in a completely new design and development effort, because the hardware configuration might be quite different. In all cases, the port must be freshly tested on the new platform to ensure that the player's experience is the same as on the original platform.

Artists

Without the art team, we would only be playing text-based games like *Hunt the Wumpus* or *Colossal Cave Adventure*. The public's expectation for the artistic experience of the game is higher than it has ever been. Games based on such transmedia franchises as *The Lord of the Rings*, *Spider-Man*, and *Star Wars* already come with built-in expectations of a visual game experience that parallels what you would see in a movie theater.

Artists apply their skills with drawing, colors, shapes, and lighting by using digital tools to create everything you see in a video game. The art elements in a game are known as art *assets*. They exist in separate files from the game code, and may be combined or compressed in some way to minimize the amount of memory they use. The art in a game may be provided on a very small scale, such as individual decals to apply to a race car in *Gran Turismo®* 7, or on a massive scale, such as rendering a planetary landscape in *Starfield*.

Animators

Animators add realism and motion to the game. Animations need to be smooth and properly scaled, taking gravity into account. An animation is made up of a series of frames. Each frame contains a specific pose, and when the frames are played together, your eyes fill in the blanks between the poses, thanks to a phenomenon known as *persistence of vision*.

Characters and creatures are believable when they are properly animated. On a large scale, they are animated to walk, run, and jump. Their movement needs to be consistent with their weight, physiology, and the local force of gravity. At a smaller level, facial expressions and body language are animated to communicate emotion.

Explosive and destructive effects provide excitement and urgency to the game. The explosion could come from the end of a gun or from a remote detonation. There will be the central core effect, such as an expanding ball of light, followed by after-effects such as smoke or a concussion wave. The explosion itself can send rocks, vehicles, or the player's opponents flying. Rocks or walls can break when a player is sent smashing into them.

Animating effects of nature help give a sense that you are part of the game environment. Leaves should spin and float as they fall, rather than drop straight down as if in a vacuum. Environmental interactions may also occur as the result of movement, such as water rippling and splashing when a character steps in or moves around in it.

Level Designers

Level designers define what goes into the various "levels" or parts of the world you explore or inhabit when playing a game. These chunks of content are defined differently based on the particular genre of the game. In a sports game, level designers might be responsible for creating authentic versions of real-life arenas or stadiums. In a racing game, they are responsible for creating authentic or fanciful tracks. For a first-person shooter (FPS) game, a real-time strategy (RTS) game, or a multiplayer online battle arena (MOBA) game, they are responsible for creating interesting and balanced maps for players to do battle. Even a vast, seamless, "open world" such as the Hyrule depicted in *The Legend of Zelda: Breath of the Wild* and *Tears of the Kingdom* is composed of regions, islands, zones, cities, and even individual quests, each of which can be thought of as a "level" for production purposes.

The level designer must make each level exciting and unique, but it should also fit into the context and theme of the overall game or world.

Sound Designers

The audio or sound team creates the audible experience of the game. Just like in the movies, a well-crafted sonic experience will make you feel as though you are part of the world. This team collaborates with designers and programmers to design sounds consistent with the visual and gameplay elements.

One of the main functions of a *sound designer* is to envision and implement an overall sound design for the game and for each level within the game. A sound design includes layered ambient noises (such as water dripping in a cave or rustling trees in a windy forest), triggered sound effects (such as the "ping" when Mario collects a coin), dialog, narration, and music. A player cannot see what is going on in the game beyond what is shown on the screen, but he can hear sounds coming from all directions. Clever game designers work closely with audio designers to create opportunities for players to use their ears to help play the game.

Game Designers

Game designers are responsible for conceiving and defining the gameplay of the game. They are storytellers, entertainers, and inventors. Their concepts are the basis of the game's worlds, characters, and lore.

The game designer also defines game mechanics that are easy to learn, remember, and access during gameplay. A *game mechanic* is anything the player can do in the game. Multiple mechanics are sometimes grouped into *systems* (like a combat system or an inventory system), and sequences of systems are sometimes referred to as *loops*.

Game design might develop from a top-down approach, where the designers start with a high-level concept and then break it down into greater detail so that the artists, sound designers, and programmers can begin to make the game. Or, game designers might use a bottom-up approach, starting with a few ideas for a fun mechanic, and then working to come up with the story and settings that tie together those mechanics and give the player reasons to keep playing.

Whichever way the game designers arrived at the game concept, they are responsible for producing the *game design document* (GDD) used by the other departments to guide how they do their parts of the project. This documentation can also be a useful resource for testers to ensure that the designers' ideas are incorporated into the game properly. Testers should use design documentation as the basis for providing a complete range of tests. From any flow charts or diagrams provided in the game design documentations, lead testers should write tests to cover all of the states and transitions possible in that particular feature or system. The same approach should be used with any screen layouts or menus documented in the game design.

Producers

Producers, often called *project managers*, work to see that the game gets done on time and within budget. Both the game developer and the game publisher may each employ their own producer. Producers develop a schedule with dates for particular production milestones (or deadlines) by which a set of tasks, goals, or *deliverables* should be finished. The deliverable provides an indication that the game is making sufficient progress and builds confidence that the game will be released on time.

In addition to developing the milestone schedule, producers may assign certain detailed tasks to specific people or job roles, along with a due date for each task. At any point in time on the schedule, the project manager should know how many people are needed, which tasks have been completed, and which tasks are in progress. Adding up all of the people at each point in time provides a staffing budget.

In addition to staffing, which translates into wages, the project manager may budget monetary expenses for equipment supplies and services. This includes new computers, hardware, and software tool licenses for programmers, artists, and testers. When some aspect of the game seems to be falling behind, producers may request more resources or figure out how to reduce the scope of the game to keep the project on schedule.

Producers rely on game testers to provide them with the best current information about the state of the game. They often determine which bugs will be fixed, in what order, and by which members of the development team.

Testers

Game testers will sometimes be identified by other titles such as test engineer, QA tester, QA analyst, or QA engineer.

There is a paradoxical relationship between programmers and testers that is based on the fact that they both want the same outcome—producing a superb game—but they have opposing roles in making that happen. Programmers want their code to run flawlessly, and will insist that their latest fix "should" work. Testers want the code to run flawlessly as well but will not believe it is true until they have conducted rigorous testing.

The tester's job is to "break" the game by finding bugs while the programmers are trying to finish the game. Testers focus on what is wrong with the game and programmers focus on making things right. This can create tension within a team and across departments on a big, multi-location project. Both developers and testers should approach their jobs with dedication and energy. However, both groups should work with each other to be better at what they do. Testers can do their jobs better by understanding how the code is designed and produced, and then using that to improve the way they write and execute tests. Programmers can continue to improve their code by learning from what kinds of problems the testers are finding, as well as by using "embedded" testers who work side-by-side with programmers to perform *unit testing* of new features or bug fixes.

The test team can be dynamic during the course of the game project. In some instances, there is little or no testing at the beginning of the project, but then many testers work on the game at the very end. This is especially true for testers who work for game publishers, because testing is how publishers make certain their investment in the game developer will result in a quality game.

The head of a game team is called the *test lead*, but they will sometimes be identified by other titles such as *QA lead* or *lead game tester*. The test lead plans and orchestrates testing activities performed over the course of the game development project. The test lead is responsible for the on-time delivery of test development and test execution results. Test activities and individual assignments are identified, planned, and adjusted as necessary during the course of the project. The test lead may be helped by one or more *primary testers*, who act as assistant managers on the team.

The test lead also establishes test procedures and standards. This includes selecting the right test tools and technologies to use for testing the game code. In many cases, test tools need to be supported or otherwise compatible with certain details of the game code. The test lead defines these "testability" requirements for the game and works with the programmers to see that they get properly implemented. On large or complex games, you may see a lead assigned to different game modes or systems, such as multiplayer, story mode, quests, dungeons, mini-games, or the character editor.

On smaller game development teams, the test lead is also responsible for doing some of the testing.

Putting It All Together

Good teams make good games. Testers are an integral and valued part of the game development team. Everyone on the team wants to deliver a good game and has one or more roles to play in making that happen, and the people on the team should be aware of the contributions and responsibilities of the other team members. Testers need to examine the individual work product delivered by each role, as well as their relationship to one another within the context of the game. Any game project documentation generated along the way—from the game design document to individual build notes—should be used by the test team to accelerate test development and improve the effectiveness of the game testing.

Before we delve into the specific phases of game testing (in Chapter 5, "The Phases of Game Quality Assurance"), let's first examine the life cycle of a typical video game development project.

THE GAME PRODUCTION CYCLE

Some games are developed in a few months. Others take several years. No matter how long projects last, each one goes through well-defined phases that have become standard across the industry. Some game projects may not incorporate all of the phases described below. It is also possible for the activities of one phase to continue while part of the team begins work on the next one:

1. Concept Development
2. Preproduction
3. Development
4. Alpha
5. Beta
6. Code Lock
7. Gold
8. Patches
9. Updates

Concept Development

Concept development is the fuzzy front end of game design. It lasts from the moment someone first comes up with a game idea until the day the game goes into preproduction. The team is very small during this period. It may consist of only the game designer, lead programmer, concept artist, and producer.

The main goal of concept development is to decide what the game is and to write this down clearly so that anyone can understand it instantly. During this phase, you decide on the major gameplay elements, create concept art to show what the game will look like, and add details to the story (if there is one).

It is important to note that no one in the games industry holds a job where they are paid simply to come up with ideas for games. "Game development teams are very collaborative, and if you can't contribute to a game's execution, you're not yet ready to contribute to its conception," writes veteran *Heroes of Might & Magic* producer David Mullich (2015).

The documents that come out of concept development are the high concept, game proposal (or "pitch doc"), and concept document.

The High Concept

The *high concept* is a one- or two-sentence description of what your game is about. It is the "hook" that makes your game exciting and sets it apart from the competition. This is sometimes referred to as an "elevator pitch," because it is a way to express an answer to the "tell me about your game" question clearly and quickly, as though you had only thirty seconds during a chance encounter on an elevator.

A strong high concept is also valuable during the development phase because it helps you to decide which features to include and which to leave out. If game development is like trying to find your way through a jungle of possibilities, the high concept is a path that has already been cleared so that you do not get lost. Any feature that does not contribute to the game's main focus is a direction you do not need to explore.

The Game Proposal ("Pitch Doc")

The *game proposal* is a short handout you speak from during pitch meetings to seek funding for the development of your game. In just a few pages, you must summarize what your game is about, why it will be successful, and how it will make money. This document covers much of the same information as the concept document, but in abbreviated form.

The Concept Document

The *concept document* is the fully developed version of the pitch doc. It is a longer document that members of the publishing team will not have time to review during a pitch meeting, but will want to read afterward to get a more detailed understanding of your game.

The concept document might contain the following sections:

- The High Concept
- Genre
- Setting
- Gameplay
- Key Features

- Target Audience
- Target Platform(s)
- Estimated Schedule, Budget, and Profit and Loss (P&L) Statement
- Competitive Analysis
- Important Team Members
- Risk Analysis

These documents should emphasize the critical points of your game and the ability of your team to deliver a quality product, on time and on budget. As a tester or test lead, you can contribute by providing time estimates for the schedule, test equipment costs for the budget, and identify test risks for the risk analysis.

Preproduction (Proof of Concept)

Preproduction is the "gearing-up" or initial planning time. Your goal is to complete the game design, create the "art bible," establish the production path, write up the project plan, and create a prototype. This phase is also used to do technical prototyping to demonstrate the feasibility of any new technology you hope to deliver. Preproduction proves both that your team can make the game and that the game is worth making.

The work products of this phase are the game design document (GDD), the art production plan, the technical design document (TDD), and the project plan, which itself is actually a suite of documents. Preproduction culminates in the delivery of the game prototype: a working piece of software that shows why the game is fun to play.

The Game Design Document (GDD)

The GDD exhaustively details everything that will happen to the player in the game. The features in this document become the requirements from which the art production plan and the technical design document are made.

During the development cycle, the GDD should always be the most current representation of everything there is to know about what the player experiences in the game. This should include complete information about the gameplay, user interface, story, characters, monsters, AI, and everything else, down to the finest detail.

Such a document, if committed to paper, would be the size of a stack of several telephone directories, impossible to maintain, read by no one, and almost instantly out of date the minute it was printed. Instead, put it on your internal network as a set of Web pages or use another robust collaborative document or Wiki solution.

Maintaining your documentation online not only has the advantage of keeping the design up to date, but also enables everyone on the team to have easy access to everything at all times. The time savings to the group over the weeks and months of development will be enormous.

Starting the Art

During preproduction, you establish the look of your game and decide how the art will be created. The designer, art director, and concept artist collaborate on developing the visual style of the game. The concept artist makes reference sheets for other artists to work from. Establishing this *art bible* early on helps to orient new artists coming on to the project and ensures the final product will have a consistent visual style throughout.

Most of this art can take the form of pencil sketches, but it is often useful in selling the game to develop a few glossy pieces that capture the high concept and pack a good visual punch.

In the early stages of the game, you should assemble a visual reference library of images that reflect the direction you want the art to take. These images can come from anywhere—magazines, travel books, movie posters, and classic paintings—as long as they are used only for guidance and do not find their way into the final product.

Establishing the Pipeline

The production *pipeline* is the process by which you go from concept to reality, from an idea in someone's head to actual figures and gameplay on the screen. For example, to create a functioning monster in an action game, you must find the most efficient way to move from a designer's specifications, to a concept sketch, to a 3D model, to a skin for the model, to animation for the figure, to applying behavior scripting to the character, to dropping it in the game and seeing how it works. All the tools you select along the way must be compatible. They must be able to "talk" with each other so that the work you do at one step can be imported to the next step, manipulated, and passed to other team members down the line.

Asset Lists, Budgets, Tasks, and Schedules

The production plan also includes the first draft of the asset list, team task lists, equipment budget, costs, and any other planning documents, including a test plan. Like the GDD, this plan must be updated and kept current throughout the life of the project.

The Technical Design Document (TDD)

The TDD sets out the way your lead programmer plans to transform the game design from words on a page to software on a machine. It establishes the technical side of the art production path, lays out the tasks of everyone involved in code development, and estimates the time to completion of those tasks. From these person-month estimates, you learn how many people you need on the project and how long they will be with you. This, in turn, has a direct effect on your budget.

In addition, the TDD specifies:

- What core tools will be used to build the game
- Which tools are already in-house
- Which tools have to be bought or created
- What hardware and software must be purchased
- What changes must be made to your organization's infrastructure (for example, storage capacity, backup capabilities, and network speed) to support development

The Project Plan

The *project plan* is the road map that tells you how you are going to build the game. It starts with the raw task lists in the TDD, establishes dependencies, adds overhead hours, and turns all that information into a real-world schedule. The final project plan is broken down into several independently maintained documents.

- **Personnel Plan:** The *personnel plan* is a spreadsheet that lists all the specific people who will be working on the game, when they will start, and their salaries. It is a road map to determining both the requirements for and costs of human labor on the project.

- **Resource Plan:** The *resource plan* calculates all the external costs of the project. It takes from the technology plan the timing of the hardware purchases to support internal personnel, and it estimates when the external costs (such as voice acting, music composition, or motion capture sessions) will be incurred.
- **Project Tracking Document:** This is where you keep track of whether you are on schedule. Some producers use project integrated management software for this (such as Jira™), some use proprietary in-house tools, and some use linked spreadsheets. The producer will track lists of individual tasks, effort (time) estimates, dependencies, and deadlines using whatever tool is most efficient to keep the team apprised of the day-to-day progress of the game.
- ***Budget:*** After applying the overhead multipliers to the manpower plan, you combine these numbers with the resource plan to derive your month-by-month cash requirements and overall *budget* for the game.
- ***Profit and Loss (P&L):*** The original *P&L* estimate was made during the concept phase. As development progresses and costs become clearer, the P&L statement must be kept current.
- ***Development Schedule:*** Many developers would rather avoid creating a firm schedule and committing to a specific release date, but you owe it to yourself and your company to do exactly that. After a release date has been set, the publisher sets their marketing, sales, and distribution teams into motion. The marketing team books advertisements that will appear in the months running up to the release date. The PR department negotiates with magazines and websites for well-timed previews and feature articles. The sales group commits to promotions with retailers. Changing the release date of the software is likely to ruin all the carefully planned efforts of these groups and result in your game selling far fewer units than it might have.
- ***Milestone Definitions:*** *Milestones* are significant points in development marked by the completion of a certain amount of work (a *deliverable*). These deliverable criteria should be concrete and very precisely defined, with language such as "Concept sketches for fifteen characters, front, side, and back" or "Weapon #1 modeled, skinned, and operational within the game with a placeholder sound effect, but

without animations or visual effects." Avoid vague deliverable criteria, such as "design 25% complete." The best milestone definitions are binary: a deliverable version either meets them or fails to, with no room for argument in between. Publishers will often instruct their QA teams to test a milestone deliverable against these criteria, which are often documented in the contract and payment schedule.

Prototype

The tangible result of preproduction is the game *prototype*. This is a working piece of software that captures on the screen the essence of what makes your game special, what sets it apart from the rest of the crowd, and what will turn it into a hit. A prototype (or *pre-Alpha* version) that contains enough art and audio content to reflect what playing the finished game will be like is sometimes called a *vertical slice*.

The prototype not only shows your vision for the game, but also establishes that your production path is working. It also gives testers their first look at the game. If the prototype includes working code, this is a good time to try out some of your tests using your test environment, which itself may be a prototype at this point in the project.

Development

Development is the longest and hardest stage of creating the game. Your development schedule may last six months to two years. Some smaller games can be designed, coded, and tested in less than six months. At the other end of the spectrum, games that are longer than two years in development run the risk of going stale, suffering personnel turnover, having features improved upon or rendered obsolete by other games, or seeing technology made useless by advances in hardware. Any of these problems can cause redesign and rework which, in turn, lead to further schedule delays.

The deceptive part of development is how long it seems at the start. You have a good plan, and it is easy to think that everything can be accomplished with time to spare. This phase of the project can be dangerously similar to summer vacation. At the beginning all you see are weeks and months stretching out in front of you, with plenty of time to do everything that you want to do. Then, as the end draws near, you wonder where all the time went and suddenly start scrambling to fit everything in.

The trick to a successful development phase is to break large tasks into smaller, more manageable tasks that are rigorously tracked. You cannot know whether you are behind on a project if you do not track the tasks. This is something you should assess daily, so you can report to your team lead or project manager whether you are on schedule or need help.

If you are an external developer working for a publisher, your progress is tracked for you in the form of milestone deliverables and dates specified in the contract. The incentive to stay on schedule is clear: if you do not meet the deadlines, you do not get paid. Well-run internal groups use the same structure. Milestones are established at the start of development and there are frequent project status meetings in which the producers get together and go over the schedule in detail.

Alpha

Alpha is the first major milestone of the game project. The majority of the game systems should be coded and implemented. Enough of the game's content should be integrated so that it feels like the game the team conceived those long months ago.

The definition of *Alpha* may vary from company to company: Generally, it is the point at which the game is mostly playable from start to finish. There might still be a few workarounds or gaps, and all the assets might not be final, but the engine, user interface, and all other major subsystems are complete and working.

As you enter Alpha, the focus starts to shift from building to finishing, from creating to polishing. This is the time to carefully consider the game features and content to decide whether any must be removed to make the schedule. This is also the time when more testers come on the team to start finding bugs.

Beta

Beta is the second major milestone of the game project. At *Beta*, all content is integrated, all development of new features or systems stops, and the only thing that should happen thereafter is bug fixing (and retesting). Stray bits of art might be upgraded, or lines of text rewritten, but the goal at this point is to stabilize the project and to eliminate as many bugs as possible before release.

The last part of Beta is often called *crunch time*. During this stage, developers and testers may stay at the office for days at a time, including weekends. While some teams may feel as though crunch time is inevitable as the final deadlines approach, the truth is that working overtime for weeks on end to the point of exhaustion makes for sloppy mistakes in the game, and those mistakes are less likely to be caught by the QA team if they have been working weeks of overtime as well.

Beta Testing

The *Beta test* phase not only gives developers valuable gameplay and balance feedback, but it is also a great way for the game team to check for defects they may have missed because there simply were not enough testers to execute large-scale multiplayer test scenarios.

Some games have a *Closed Beta* testing period, where volunteer testers are either randomly chosen or hand-picked based on information they send in when they request to participate. Depending on the project plan or the results from the Closed Beta, there may be a subsequent *Open Beta* period, which tests the game at an even larger scale. Other rounds of Beta testing may be defined with their own specific objectives, such as testing how well a massively multiplayer online role-playing game (MMORPG) performs with a fully loaded world server.

Compliance Testing

If yours is a console game that is subject to the approval of the console manufacturer, the final weeks of Beta will include submissions to that company so its testers can verify that the game meets their own quality standards.

A PC game can be sent to an outside testing firm for compatibility testing. This should uncover any pieces of hardware, or combinations of hardware, that keep the game from working properly.

A mobile game may need approval from handset manufacturers, wireless carriers, and digital storefronts like the App Store or Google Play. The handset manufacturers may be concerned about the game's interoperation with a phone's built-in features, while the carriers may want to be confident that the game will not disrupt service on their network.

Code Lock

At the end of Beta, you are likely to be in *code lock*, when all the work is done and the preparation of a master candidate version begins. In the case of disk-based games, the disks must be tested before they are sent to the manufacturer. In the case of digital-only games, a final version is prepared with all the specifications required for it to be uploaded to the release server. The only changes allowed to the code base are those that specifically address any "show-stopper bugs" that last-minute testing may find.

Gold

This is the last major milestone in the game development life cycle. The game's released to manufacture (RTM) version has been thoroughly tested and found to be acceptable when measured against all release criteria, whether from the publisher, the platform owner, or the digital storefront. This version, the one distributed to the playing public, is known as the *Gold* version, and finishing the game is often known as "going Gold." This term dates back to when all game software was manufactured in quantities and sold in stores. The version sent to be mass produced on CD-ROMs was known as the *Gold Master*, and versions created during the code lock were known as *Gold Master Candidates* (GMCs).

The game is finished. (Or is it?) You can finally celebrate. (Or can you?)

Patches

For better or worse, it seems today as though almost every game gets a bug-fixing patch after its initial release. (We will explore why that is, despite months of game testing, later in the book.)

Immediately after the game's release, these first *patches* tend to focus on repairs to outstanding bugs, as well as any new bugs that may have been found by players.

Updates

An *update* is different from a patch. An update contains additional content created to expand the original game. Many game companies conduct seasonal or holiday events for their subscribers, which could include special limited-time missions, crafting options, or item drops. Many competitive games will introduce new cards, new playable characters, or other new gameplay elements that will affect the *balance* of the game. These are delivered in updates, with further updates sometimes necessary to fine-tune balance.

In any case, an update is a mini-project and needs to be handled like one, with testing, milestones, and all the other factors associated with good software development and testing.

READY, TESTER ONE?

Now that you understand the role of testing in the game development process, in the next chapter we will examine some of the qualities and discipline necessary to become a professional game tester.

EXERCISES

1. The number of testers on a video game project will change over the course of development. Will there be more testers on the team toward the beginning or toward the end?
2. What is the main difference between a patch and an update?
3. Why is it usually a bad idea to print out a game design document (GDD)?
4. Which person is responsible for planning and orchestrating testing activities performed over the course of a game development project?
5. What is the purpose of a game prototype?
6. How can game testers contribute to the management of a game development project?
7. What are the three major milestones of a game development project, and why is it important to have them precisely defined?

REFERENCE

Mullich, David. 2015. "Sorry, There Is No 'Idea Guy' Position In The Game Industry." Last modified November 23, 2015. *https://davidmullich.com/2015/11/23/sorry-there-is-no-idea-guy-position-in-the-game-industry*.

CHAPTER 2

The Basics of Game Testing

Developers do not fully test their own games. They do not have time to, and even if they did, it is not a good idea, for reasons that will become clear later in the book.

At the beginning of the video game era, the programmer of the game was also its artist, designer, and tester. Even though games were very small (the size of a modern email message), the programmers spent most of their time designing and programming; they spent very little time testing. If they did any testing, it was based on their own assumptions about how players would play the game. The following section illustrates the type of problem these assumptions can create.

The Player Will Always Surprise You

The programmer of *Astrosmash*, an arcade-style shooting game released for the Intellivision® console system in 1981, made an assumption when he designed the game that no player would ever score ten million points. As a result, he did not bother to write code that would check for score overflowing. He read over his own code and based on his own assumptions, it seemed to work fine. It was a fun game, its graphics were breathtaking (for the time), and the game went on to become one of the bestsellers on the Intellivision platform.

Weeks after the game was released, however, a handful of customers began to call the game's publisher, Mattel Electronics, with an odd complaint: when they scored more than 9,999,999 points, the score displayed negative numbers, letters, and symbols. This occurred in spite of the promise of "unlimited scoring potential" in the game's marketing materials. The problem was made worse by the fact that the Intellivision console had a feature that allowed players to slow the game down, making it much easier to earn extremely high scores. John Sohl, the programmer, learned an early lesson about video game development: *the player will always surprise you*.

This story demonstrates that video game testing is best done by testers who are (a) professional, (b) objective, and (c) separated—either physically or functionally—from the game's development team. That distance and objectivity allows testers to think independently of the developers, to function as players, and to figure out new and interesting ways to "break" the game. When a tester discovers something wrong with the game program, they have found a *defect* (or *bug*) in the software. A game bug can be defined as a deviation from the player experience intended by the game designers.

Don't Say "Glitch!"

Using the term "bug" to describe a software defect dates back to the earliest days of programming, when the heat and light of primitive computers would attract insects inside the machine itself, which sometimes led to malfunctions. "Glitch," however, comes from the field of television broadcasting. A visible error when playing back a video tape was often called a glitch.

If you want to sound professional when talking about a mistake in a video game, call it a bug or a defect.

NOT ALL PLAYERS ARE ALIKE

Game developers work hard to create an engaging, well-designed experience for their players. But different people play video games for different reasons, and different players play with different styles. In recent years, the study of *player types* has been fruitful for game designers and marketers, because as the video game industry has grown, matured, and become more diverse, so has its audience. Market research firm Quantic Foundry published a survey of player motivation that shows that different people are play games for a range of different feelings, among them:

- The thrill of daunting challenges
- The joy of discovering new places or finding hidden things
- The satisfaction of completing a task or collection
- The excitement of destroying things and defeating enemies
- The rewards of executing a careful plan
- The drama of a compelling story

(You can take this survey and discover what type of player you are at *https://apps.quanticfoundry.com/surveys/start/gamerprofile*.)

Not every game is intended for every type of player. Game designers tend to design for a specific audience, and try to craft an engaging experience for their players. Those players tend to "color within the lines," as they do not want anything to break their sense of immersion while playing the game. It is the job of the game tester to "color outside the lines."

BLACK BOX TESTING

Almost all game testing is *black box* testing, which is testing done from outside the application. Typically, no knowledge of, or access to, the source code of the game is granted to the game tester. Testers do not find defects by reading the game code. They try to find bugs by using the same input devices available to the player, be it a mouse, a keyboard, a console gamepad, a touch screen, a motion sensor, or a plastic guitar controller. Black box testing is the most cost-effective way to test the extremely complex network of systems and modules that even the simplest video game represents.

Figure 2.1 illustrates some of the various inputs you can provide to a video game and the outputs you can receive back. The most basic of inputs are position and control data in the form of button presses and cursor movements. Players can also input inputs vector and force data from accelerometers, cameras, or other special controllers. Audio input can come from microphones fitted in headsets or attached to a game controller. Input from other players can come from a second controller, a local network, or the Internet. Finally, stored data such as saved games and options settings can be called up as input from memory cards or a hard drive.

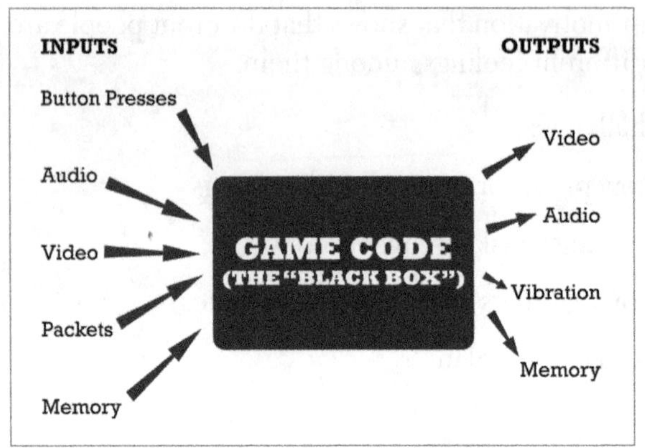

FIGURE 2.1 Black box testing: planning inputs and examining outputs

Once some or all of these types of input are received by the game program, it reacts in interesting ways and produces such output as video, audio, vibration (via force feedback devices), and data saved to memory cards or hard drives.

The input path of a video game is not one-way, however. It is a feedback loop, where the player and the game are constantly reacting to each other. Players do not receive output from a game and stop playing. They constantly alter and adjust their input moment by moment based on what they see, feel, and hear in the game. The game, in turn, makes similar adjustments in its outputs based on the inputs it receives from the player. Figure 2.2 illustrates this loop.

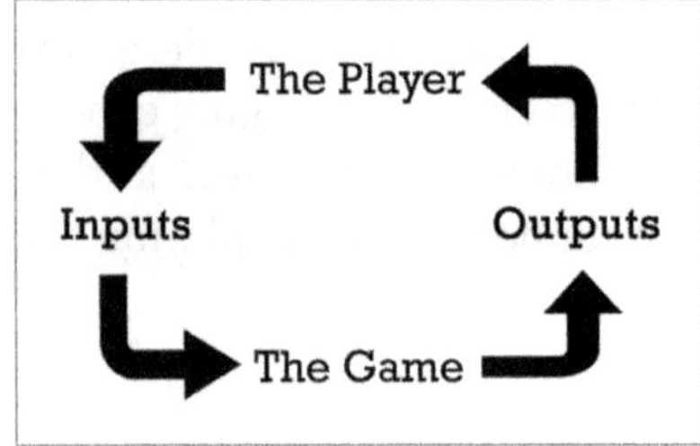

FIGURE 2.2 The player's feedback loop adjusts to the game's input, and vice versa.

If the feedback received by the player was entirely predictable all the time, the game would be no fun. Nor would it be fun if the feedback received by the player was entirely random all the time. Instead, feedback from games should be just random enough to *seem* unpredictable. It is the unpredictability of the feedback look that makes games fun. Because the code is designed to surprise the player and the player will always surprise the programmer, black box testing allows testers to think and behave like players do.

WHITE BOX TESTING

In contrast to black box testing, *white box* testing gives the tester opportunities to exercise the source code directly in ways that no player ever could. It can be a daunting challenge for the white box tester to read a piece of game code and then try to predict every single interaction it will have with every other bit of code, and whether the programmer has accounted for every combination and order of inputs possible. Testing a game using only white box methods is also extremely difficult because it is nearly impossible to account for the complexity of the player feedback loop. There are, however, situations in which white box testing is more practical and necessary than black box testing. These may include:

- Tests performed by developers prior to submitting new code for integration with the rest of the game.
- Testing code modules that will become part of a reusable library across multiple games or platforms.
- Testing code methods or functions that are essential parts of a game engine or middleware product.
- Testing code modules within your game that might be used by third-party developers or "modders" who could expand or modify the behavior of your game to their own liking.
- Testing low-level routines that your game uses to support specific functions in the newest hardware components, such as graphics cards or audio processors.

In performing white box tests, you execute specific modules and the various paths that the code can follow when you use the module in various ways. Test inputs are determined by the types and values of data that can be passed to the code. Results are checked by examining values returned by the module, global variables that are affected by the module, and local variables as they are processed within the module. For an example of white box testing, consider the `TeamName` routine from *Castle Wolfenstein: Enemy Territory*:

```
const char *TeamName(int team)    {
    if (team==TEAM_AXIS)
        return "RED";
    else if (team==TEAM_ALLIES)
        return "BLUE";
    else if (team==TEAM_SPECTATOR)
        return "SPECTATOR";
    return "FREE";
}
```

(id Software 2012)

Four white box tests are required for this module to test the proper behavior of each line of code within the module. The first test would be to call the `TeamName` function with the parameter `TEAM_AXIS` and then check that the string "`RED`" is returned. Second, pass the value of `TEAM_ALLIES` and check that "`BLUE`" is returned. Third, pass `TEAM_SPECTATOR` and check that "`SPECTATOR`" is returned. Finally, pass some other value such as `TEAM_NONE`, which makes sure that "`FREE`" is returned. Together these tests not only exercise each line of code at least once, they also test the behavior of both the "true" and "false" branches of each if statement.

This short exercise illustrates some of the key differences between a white box testing approach and a black box approach:

- Black box testing should test all of the different ways you could choose a test value from within the game, such as different menus and buttons. White box testing requires you to pass that value to the routine in one form: its actual symbolic value within the code.
- By looking into the module, white box testing reveals all of the possible values that can be provided to and processed by the module being tested. This information might not be obvious from the product requirements and feature descriptions that drive black box testing.
- Black box testing relies on a consistent configuration of the game and its operating environment to produce repeatable results. White box testing relies only on the interface to the module being tested and is concerned only about external files when processing streams, file systems, or global variables.

YOUR FIRST TEST SUITE

At home, you play games to have fun. You get to choose what to play, when to play, and how to play. Testing games can still be fun, but you have fewer choices about what, when, and how to play. Everything you do as a tester is for a purpose: either to explore an area of the game, to check that a specific system is working properly, or to look for a particular kind of problem.

Your job as a tester begins by running a series of specific tests that are assigned to you. Many of these tests are very specific and consist of step-by-step instructions. These rely on your keen observations and attention to detail. These types of *directed tests* may have been developed by the lead tester or by the developers themselves, but are always based on the designer's specifications. They may come in different formats like spreadsheets or paper check lists.

Below is short example of a directed user interface (UI) test of the character selection portion of the Edit Trooper UI in *Song Summoner: The Unsung Heroes—Encore* mobile game (Figure 2.3).

FIGURE 2.3 *Song Summoner* Edit Trooper selection screen showing the initial state of the character gallery

This checklist is a series of instructions asking you specific questions about certain details of the UI. In a common test setting, if any of the features or values described below the list is not present, then that part of the test would fail. You would report that failure by writing a bug report.

1. Enter a town and tap the "Edit Trooper" icon to enter the Edit Trooper screen.
   - ❑ Check that the main character's picture appears first in the list, with his name (Ziggy) above his image. Also check that the image frame and background reflect his current status in the game (in this case, "Gold").

- ❑ Check that the circular control within the slider bar is positioned to the far left, the number "1" appears to the left of the bar, and the number at the right of the slider bar equals the number of characters you have acquired during the game (83 in this example).
- ❑ Check that Ziggy's character class "Capable Conductor" appears in the upper left of the box in the lower left corner of the screen.
- ❑ Check that Ziggy's stats for Movement, Range, HP, and SP are correct.
- ❑ Check that a miniature version of Ziggy's avatar appears to the right of his stats, with a small musical note at his right.

2. Scroll one character over by swiping the screen from right to left.
   - ❑ Check that the second character's picture appears first in the list, with his name above his image. Also check that the image frame and background reflect his current status in the game.
   - ❑ Check that the circular control within the slider bar moved slightly to the right and that the numbering on the ends of the bar has not changed.
   - ❑ Check that the newly selected character's class appears in the upper left of the box in the lower left corner of the screen.
   - ❑ Check that the newly selected character's stats for Movement, Range, HP, and SP are correct.
   - ❑ Check that a miniature version of the new character's avatar appears to the right of his or her states, with a small musical note to the right.

3. Scroll one character to the right by holding your finger on the circular control within the slider bar and dragging it slightly to the right until the third character's image is in the center of the screen.
   - ❑ Perform the same checks that you made in Step 2, but this time for the proper display of the third character's information.

4. Scroll one character to the left by swiping the screen from left to right.
   - ❑ Perform the same checks as you made in Step 2.

5. Scroll one character to the left by holding your finger on the circular control within the slider bar and dragging it slightly to the right until the first character's image is in focus.
   - ❑ Perform the same checks as you made in Step 1.

6. Scroll all the way to the end of the trooper list (see Figure 2.4) by swiping from right to left (multiple times if necessary).
   - ❑ Check that the last character's picture appears first in the list, with his name above his image. Also check that the image frame and background reflect the character's current status in the game.
   - ❑ Check that the circular control within the slider bar is positioned to the far right and that the number "1" appears to the left of the bar and the number at the right of the slider bar equals the number of characters you have acquired during the game.
   - ❑ Check that the last character's class appears in the upper left of the box in the lower left corner of the screen.
   - ❑ Check that the last character's stats for Movement, Range, HP, and SP are correct.
   - ❑ Check that a miniature version of the last character's avatar appears to the right of his or her stats, with a small musical note to the right.

FIGURE 2.4 *Song Summoner* Edit Trooper selection screen showing the end of the character gallery

7. Scroll all the way back to the beginning of the list by holding your finger on the circular control within the slider bar and dragging it all the way to the left until the first character's image is in focus.
   - ❑ Perform the same checks that you made in Step 1.

8. Scroll all the way to the end of the list by holding your finger on the circular control within the slider bar and dragging it all the way to the right until the last character's image is in focus.
   - ❑ Perform the same checks that you made in Step 6.

9. Scroll all the way to the beginning of the trooper list by swiping from left to right (multiple times if necessary).
   - ❑ Perform the same checks that you make in Step 1.

NOTE *This example, like many of examples from specific games that appear throughout this book, are meant only to demonstrate an underlying concept that should be applicable across all games. You do not need to have played (or have seen) any of the specific games mentioned in these examples. Instead, think of a game you have played that has a similar module or feature. Could you write this UI testing checklist on that last game you played with a character selection module?*

In directed testing, you are given specific operations to perform and details to check at each step. This can become tedious over the course of a long test, especially when doing many of these tests one after the other. To keep subtle problems from slipping past you, maintain concentration and treat each item as if it is the first time you have seen it.

WRITING TEST CASES AND TEST SUITES

A single test performed to answer a single question is a *test case*; a collection of test cases is a *test suite* (and sometimes a *checklist* or *matrix*). The checklist we discussed in the last section is an example of a test suite.

The lead tester, primary tester, or any other tester tasked with creating test suites should draft these documents prior to the distribution of a new version of the game software, which is commonly referred to as a *build*. Depending upon the complexity of the game, the composition of the development and test teams, and the phase of the project schedule, the programmers might release builds semi-monthly, weekly, daily, or sometimes more than once a day. Each tester will take their assigned test suites and perform them on the new build. Any defects not already documented in the *bug database* should be reported as new bugs. (In the next chapter, we will discuss writing bug reports and tracking them with the bug database.)

In its simplest form, a test suite is a series of incremental steps that the tester can perform sequentially. Later chapters in this book discuss in detail the careful design of test cases and suites through such methods as combinatorial tables and test flow diagrams. For the purposes of this discussion, consider a short test suite you might execute on the classic PC game *Minesweeper*. A portion of this suite is shown in Figure 2.5. (You will find a sample test suite in Appendix B of this book.)

Step	Pass	Fail	Comments
1. Launch Minesweeper			
2. Musical tone plays?			
3. Visible menu options are Game and Help?			
4. Right Number (time elapsed) displayed as 0?			
5. Left Number (bombs left) displayed is 10?			
6. Click Game on the menu and choose Exit.			
7. Game closes?			
8. Re-launch Minesweeper.			
9. Choose Game > Options > Custom			
10. Enter 0 in the Height box			
11. 0 accepted as input?			
12. Click OK.			
13. Error message appears?			
14. Click OK again.			
15. Game grid 9 rows high?			
16. Game grid 9 columns wide (unchanged)?			
17. Choose Game > Options > Custom			
18. Enter 999 in the Height box			
19. 999 Accepted as input?			
20. Click OK.			
21. Playing grid 24 rows high?			
22. Playing grid 9 columns wide (unchanged)?			

FIGURE 2.5 A portion of a test suite for *Minesweeper*

This is a very small portion of a very simple test suite for a very small and simple game. The first section (Steps 1-7) tests launching the game, ensuring that the default display is correct, and exiting. Each step either gives the tester one incremental instruction or asks the tester one simple question. Ideally, these questions are binary and unambiguous. The tester performs each test case and records the result.

Because the tester may observe results that the test designer had not planned for, the "Comments" field allows the tester to elaborate on a "yes/

no" answer, if necessary. The lead or primary tester who receives the completed test suite can then scan the "Comments" field and make adjustments to the test suite as needed for the next build.

Where possible, the questions in the test suite should be written in such a way that a "yes" answer indicates a "pass" condition: the software is working as designed and no defect is observed. "No" answers, in turn, should indicate that there is a problem and a defect should be reported. Doing this is intuitive, because we tend to group "yes" and "pass" (both positives) together in our minds the same way we group "no" and "fail." Further, by grouping all passes in the same column, the completed test suite can easily be scanned by both the tester and test managers to determine quickly whether there were any failures. A clean test suite will have all the checks in the "Pass" column.

For example, consider a test case covering the display of a *tool tip*, a small pop-up window with instructional text that is incorporated into many interfaces. A fundamental test case would be to determine whether the tool tip text contains any typographical errors. The most intuitive question to ask in the test case is: `Does the text contain any typographical errors?`

The problem with this question is that a "Pass" (no typos, hence no bugs) would be recorded as a "no." It would be very easy for a hurried (or tired) tester to mistakenly mark the "Fail" column. It is far better to express the question so that a "yes" answer indicates a "pass" condition: `Is the text free of typographical errors?`

As you can see, directed testing using test suites is structured and methodical. Other test assignments involve more open-ended directives, and might be in checklist or outline form. These tests rely more on your own individual game knowledge, experience, and skills. As you grow in the position and earn the confidence of your test lead and the other members of your team, you may be asked to do more *exploratory* testing, which is discussed in more detail in later chapters.

THE LIFE CYCLE OF A BUILD

Players wait for new games; testers wait for new builds. Understanding the test results from each build is how all the stakeholders in the project—from QA to the producer to the publisher—measure the game's progress toward release.

A basic game testing process consists of the following steps:

1. **Plan and design the test.** Although much of this is done early in the project planning phase, planning and design should be revisited with every build. What has changed in the design since the last build? What additional test cases should be added? What new configurations will the game support? What features have been cut? The scope of testing should ensure that no new issues were introduced during process of fixing bugs prior to this release.
2. **Prepare for testing.** Code, tests, documents, and the test environment are updated by their respective owners and aligned with one another. By this time, the development team should have marked the bugs fixed for this build in the defect database so that the QA team can verify those fixes and close those bugs.
3. **Perform the test.** Run the test suites against the new build. If you find a defect, test "around" the bug (discussed below) to make certain you have all the details necessary to write as specific and concise a bug report as possible. The more research you do in this step, the more specific and useful the bug report will be.
4. **Report the results.** Log the completed test suite and report any defects you found in the bug database.
5. **Repair the bug.** The test team participates in this step by being available to discuss the bug with the development team and to provide any directed testing a programmer might require in order to track the down the defect. Once the programmer thinks they have fixed the bug, they will compile a new build to submit for the next round of testing.
6. **Return to Step 1 and re-test.** With new bug reports and new test results comes a new build. The test team starts the process over again. The first step in re-testing will be to do what is known as *regression testing*, which means testing those bugs which the developer claims to have fixed in the new build.

These steps not only apply to black box testing; they also describe white box testing, configuration testing, compatibility testing, and any other type of QA. These steps are identical no matter what their scale. If you substitute the word "game" or "project" for the word "build" in the preceding steps, you will see that they can also apply to the entire game, a phase of development (such as Alpha or Beta), or an individual module or feature within a build. In this manner, the software testing process is somewhat fractal: the smaller system is structurally identical to the larger system, and vice versa.

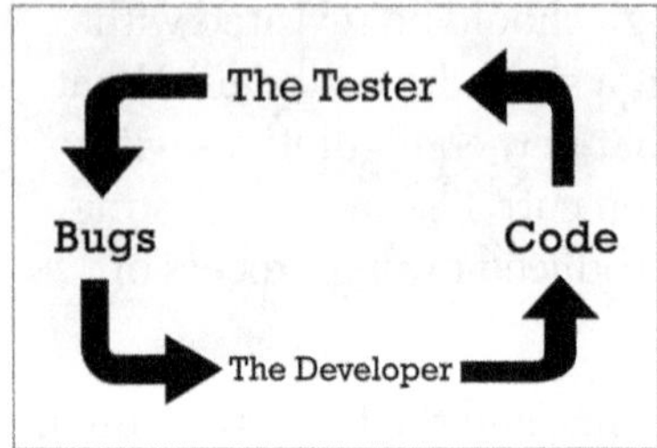

FIGURE 2.6 The testing process feedback loop

The Testing Feedback Loop

As illustrated in Figure 2.6, the testing process itself is a feedback loop between the tester and developer. The tester plans and executes tests on the code, then reports the bugs to the developer, who fixes them and compiles a new build, which the tester plans and executes tests on, and so forth.

The Essence of a Bug Report

At some point, whether through directed testing or exploratory testing, you will find a bug. You will then have to write a report on that bug, and these reports are tracked in the project's bug database.

We will discuss good bug report writing in the next chapter, but for now let's examine the purpose of the report. Bug reports must be written as step-by-step instructions so that developers can recreate the bug, understand what problem in the software is creating the bug, and correct the problem so that if the tester follows the same steps in a new build, the bug no longer occurs.

Because everyone on the team is always busy, it is the job of the tester when writing a bug report to do a combination of research and detective work to reduce to a minimum the number of discrete actions required to get the bug to occur. This is why testers should both record their gameplay and take notes as they test, so that they can be sure that they are reporting the best path to the "root cause" of the defect.

Testing "Around" a Bug

An old saying in carpentry is "measure twice, cut once." Good game testers thoroughly investigate a defect before they write it up, anticipating any questions the development team might have.

Before you begin to write a defect report, ask yourself some questions:

1. Is this the only location or level where the bug could occur?
2. Does the bug occur while using other characters or units?

3. Does the bug occur in other game modes (for example, multiplayer as well as single-player, skirmish as well as campaign)?
4. Can you eliminate any steps along the path to reproducing the bug?
5. Does the bug occur across all platforms, or is it specific to, say, the PS5 build?
6. Is the bug machine-specific? (For example, does it occur only on PCs with a certain hardware configuration?)

These are types of questions you will be asked by your lead tester, the project manager, or the developers. Try to adopt the habit of preempting such questions by performing some quick additional tests before you write the bug report. Test to see whether the defect occurs in other areas. Test to determine whether the bug happens when you choose a different character. Test to check which other game modes may contain the issue. This practice is known as testing "around" the bug.

Once you are satisfied that you have anticipated any questions that the development team might ask, and you have all your facts ready, then you are finally prepared to write the bug report, which we will discuss in the next chapter.

Interview: Brent Samul

More people are playing games than ever before. The population of game players has grown exponentially over the last two decades and has become more demographically diverse. Players are different from each other, have different levels of experience with games, and play games for a range of reasons. Some players want a competitive experience, some want an immersive experience, and some want a gentle distraction.

The pool of game testers in any organization is always less diverse than the player base of the game they are testing. Game testers are professionals, they have skills in manipulating software interfaces, and they are generally (but not necessarily) experienced game players. It is likely that if your job is creating video games, then you have spent years playing many video games. But not every player is like you.

Brent Samul, QA Lead for developer Mobile Deluxe, puts it this way: "The biggest difference when testing for mobile is your audience. With mobile, you have such a broad spectrum of users. Having played games for so long myself, it can sometimes be really easy to overlook things that someone who doesn't have so much experience in games would get stuck on or confused about."

It is a big job. "With mobile, we have the ability to constantly update and add or remove features from our games. There are always multiple things to test for with all the different configurations of smartphones and tablets that people have today," Mr. Samul says.

Although testers should write bugs against the design specification, the authors of that specification are not omniscient. As the games on every platform become more complex, it is the testers' job to advocate for the players—all players—in their bug writing. (Permission: Brent Samul)

NOT ALL TESTERS ARE ALIKE

Just as different players play games for different reasons, different members of the test team may be more or less suited for different aspects of game testing. Not every tester will run the same test in the same way. Just as a good superhero team is made up of heroes with different powers and abilities, a good game testing team is made up of different people with a diverse mix of attitudes and aptitudes. One aspect of the Myers-Briggs® personality framework suggests that people are either Judgers or Perceivers (Myers & Briggs Foundation 2024).

Judgers require a very structured, ordered, and predictable environment to be happy. If judgers are working in an unorganized environment, they will either try to organize it or they will complain that things are a mess, nothing is in its place, or that the disorganized workplace environment affects their productivity. Judgers thrive in unionized or highly regulated environments. Judgers work first and play later.

Perceivers are more relaxed and comfortable with disorder. Perceivers focus on the experience and prefer things to unfold as they will. Perceivers do not like to limit their options and thrive in dynamic, ever-changing work environments. Perceivers can work in a mess; in fact, they sometimes prefer to work in chaos, as they feel it stimulates creative thinking. Perceivers love to play, therefore if work is playful and unconventional, they are happy.

Your tendency to identify with one of these types will manifest itself in the way you approach testing, and the kinds of defects you find. For example, a Judger is good at following step-by-step instructions, running through a long list of test cases, and finding problems in the game text, the user manual, and anywhere the game is inconsistent with historical facts. The Perceiver tends to wander around the game, come up with unusual situations to test, report problems with playability, and comment on the overall game experience.

Conversely, there are things Judgers and Perceivers might not do particularly well. A Judger would perhaps not do steps or notice problems that are not documented in the written tests. A Perceiver could miss seeing problems when running a series of repetitive tests. Although testing without written tests provides more freedom, Perceivers might not always have good notes about how they got a bug to occur.

You most likely feel some affinity with both types, but may identify more with one than the other. Do not treat that as a limitation. Use that knowledge to become more aware of areas where you can improve so that you can find more bugs in the games you test. Your goal should be to use either attitude at the appropriate time and for the right purpose. When you see a bug that someone else found and it makes you think "Wow! I never would have tried that," then go and talk to that person. Make sure you share your own bug stories as well.

Table 2.1 shows some of the ways that each personality type affects the kinds of bugs testers will find and what kinds of testing are best used to find them.

TABLE 2.1 Tester Personality Comparison

Judger	Perceiver
Run the tests for...	Find a way to...
Conventional game playing	Unconventional game playing
Repetitive testing	Testing variety
User manual, script testing	Gameplay, usability testing
Factual accuracy of the game	Realistic experience of the game
Step-by-step or checklist testing	Open-ended or outline-based testing
May rely too much on test details to see defect	May stray from the original test purpose
Concerned about game quality	Concerned about game quality

EXERCISES

1. Identify each of the following as Judger (J) or Perceiver (P) behaviors:
   a. Noticed a misspelling in the user manual
   b. Created a character with all skills set to zero just to see what would happen
   c. Reported that an F1 racing car is painted in last season's livery
   d. Found a way to get the player character to skate off the map
2. What is the difference between a test case and a test suite?
3. True or False? The lead programmer is responsible for drafting test suites.
4. True or False? White box testing gives the tester access to the source code of the game.
5. How does a test suite help to identify bugs?
6. Which of the following is NOT a professional term to describe an unexpected result when playing a game?
   a. a software defect
   b. a glitch
   c. a bug

REFERENCES

id Software. 2012. "Enemy-Territory/scr/game/g_items.c." Accessed January 7 2024. *https://github.com/id-Software/Enemy-Territory/blob/master/src/game/g_items.c.*

Myers & Briggs Foundation. 2024. "Myers-Briggs® Overview." Accessed January 3, 2024. *https://www.myersbriggs.org/my-mbti-personality-type/myers-briggs-overview/*

CHAPTER 3

Bug Report Writing and Defect Tracking

HOW TO WRITE A BUG REPORT

One of the most important skills a video game tester can learn is good writing. A defect can be fixed only if it is communicated clearly and effectively.

Good bug report writing gets the development team to understand the bug you have found. In the report you will explain **what** the problem you found is, **where** in the game it is located, **when** (and how often) it happens, and most importantly **how** anyone on the team can recreate the bug. The developers are not the only people who will read your bug report, however. Your audience could include:

- The lead tester or primary tester, who might wish to review the bug before they "open" it to make it active in the bug database. (We will explore the bug database later in this chapter.)
- The producer, who will read the open bug and assign it to the appropriate member of the development team.
- Marketing and other executives, who might be asked to determine the possible business impact of fixing (or not fixing) the bug.
- Third parties, such as middleware developers, who could be asked to review a bug that is possibly related to a product they license to the project team.

- Customer service representatives or community managers, who might be asked for workarounds for the bug.
- Other game testers, who will reproduce the steps if they are asked to verify a fix in a during regression testing of a future build.

Because you never know exactly who will be reading your bug report, you must always write in as clear, objective, and dispassionate a manner as possible. You cannot assume that everyone reading your bug report will be as familiar with the game as you are. Testers spend more time in the game—exploring every hidden path, closely examining each asset—than almost anyone else on the entire project team. A well-written bug report will give a reader who is not familiar with the game a good sense of the type and severity of the defect it describes.

Just the Facts, Please

The truth is that defects often stress and annoy development teams, especially toward the end of the project during "crunch time." Each new bug added to the database means that there is more work still to be done. An average-sized project can have hundreds or thousands of defects reported before it is completed. Developers can feel overwhelmed and might, in turn, get hostile if they feel their time is being wasted by frivolous or arbitrary bugs. That is why good bug writing is fact-based and unbiased. Consider the following example:

```
The guard's hat would look better if it was green.
```

This is neither a defect nor a fact. It is an unsolicited opinion about visual design. There are forums for such opinions (discussions with the lead tester, team meetings, or play testing feedback), but the bug database is not one of them.

Know Your Role!

As a QA tester, your job is to test the game, not to second-guess the game designers. Your personal opinions about the game or any of its assets or features or systems does not belong in the bug database.

A common complaint in some games is that computer-controlled enemies appear too easy to beat or that the independent unit pathfinding is inefficient. Artificial Intelligence (AI) is the commonly used catch-all term that means any opponents or NPCs controlled by the game code (even if that code is not really using true AI techniques, such as deep learning). Consider the following example:

```
The AI is weak.
```

This could indeed be a commonly held opinion, but it is simply that—an opinion. A much better way to convey the same information would be to isolate and describe a specific example of faulty "AI behavior" and write up that specific defect. By revising such statements from broad generalizations to specific facts, you can turn them into defects that have a good chance of being addressed.

Bug reports are complex, and busy team members (even your fellow testers) do not have time to read through the full text of the bug report each time they encounter it. For this reason, it has been common practice for decades to include two key fields in every bug report, the "headline" of the bug, or its *brief description*, and the steps to reproduce the bugs, or its *full description*.

The Brief Description

The brief description field is used as a quick reference to identify the bug. This should not be a cute nickname, but a one-sentence description that allows team members to identify and discuss the defect without having to read the longer Full Description each time. Consider the following example:

```
Crash to desktop.
```

This is not a complete sentence, nor is it specific enough to be a brief description. It could apply to any one of dozens of defects in the database. The brief description must be brief enough to be read easily and quickly, but long enough to describe the bug. Consider the following example:

```
The saving system is broken.
```

This is a complete sentence, but it is not specific enough. What did the tester experience? Did the game not save? Did a saved game not load? Does saving cause a crash? Consider the following example:

```
Crash to desktop when choosing "Options" from Main Menu.
```

This is a complete sentence, and it is specific enough so that anyone reading it will have some idea of the location and severity of the defect. This well-written brief description follows the "PAL" format, which is a very good method of composing a bug headline. Write a sentence that describes a specific "**P**roblem" caused by a specific "**A**ction" in a specific "**L**ocation." Consider the following example:

```
Game crashed after I killed all the throne room guards and
doubled back through the level to get all the pick-ups and
killed the first respawned guard.
```

This is a run-on sentence that contains far too much detail. A good way to revise it might be as follows:

```
Game crashes after killing respawned guards in throne room.
```

Write the Story Before the Headline

Write your full description first, then write the brief description. Spend some time polishing the full description to help you understand the most important details to include in the brief description.

The Full Description

If the brief description is the headline of the bug report, then the full description provides the whole story. Rather than being a prose discussion of the defect, the full description should be written as a series of simple instructions so that anyone can follow the steps and reproduce the bug. Like a cooking recipe—or computer code, for that matter—the steps should be written in the second person imperative mood, as though you were telling someone what to do. The last step should be a sentence (or two) describing the bad result. For example:

```
1. Launch the game.

2. Watch the animated logos. Do not press ESC to skip through
   them.

-> Notice the bad strobing effect at the end of the developer's
   logo.
```

Here, "notice" instructs the reader to see the bad result of following the steps. The tester is probably safe to assume that the designers do not want players to be disturbed by a jarring (and possibly harmful) strobing effect, especially before the actual game has even begun. So the final step, set off with an arrow bullet, asks the reader to notice or observe the problem.

The fewer steps it takes to reproduce the defect, the better it is for everyone on the development team. Using fewer words per step saves time. Remember Brad Pitt's warning to Matt Damon in the Hollywood film *Ocean's Eleven* (2001): "Don't use seven words when four will do." Time is a precious resource when developing a video game. The less time it takes a programmer to read, reproduce, and understand the bug, the more time they have to fix it. Consider the following example:

```
1. Launch game.

2. Choose multiplayer.

3. Choose skirmish.

4. Choose "Sorrowful Shoals" map.

5. Choose two players.

6. Start match.

-> Map fails to load.
```

These are very clear steps, but for the sake of brevity, they could be shortened to the following:

```
1. Start a two-player skirmish match on "Sorrowful Shoals."
```

Sometimes, however, you need several steps. The following bug describes a problem with a power-up called "mugging" in the 2001 PC real-time strategy game *Battle Realms*. This power-up was created by the designers to allow the Swordsman unit to steal any other power-up from any other unit. The designers did not intend for the mugging power-up to crash the game itself. (As this bug requires two testers to coordinate efforts, there are instructions for both a Serpent clan tester and an opponent tester.)

```
1. Create a game against one human player. Choose Serpent clan.

2. Send a Swordsman into a Thieves Guild to get the Mugging
   power-up.

3. Have your opponent create any unit and give that unit any
   power-up.

4. Have your Swordsman meet the other player's unit somewhere
   neutral on the map.

5. Activate the Mugging power-up.

6. Attack your opponent's unit.

-> Crash to desktop when Swordsman strikes.
```

This might seem like many steps, but it is the quickest way to reproduce the bug. Every step is important to isolate the behavior of the mugging code. Even a small detail, like meeting in a neutral place on the game map, can be very important, because meeting in occupied territory might bring allied units from one side or the other into the fight, and the test might be impossible to perform. Both testers would need to start the steps again, wasting valuable time.

Good bug writing is precise *yet* concise.

Great Expectations

Sometimes the defect itself will not be obvious from the steps in the full description. Because the steps produce a result that deviates from player expectations, but does not produce a crash or other severe or obvious symptoms, it is sometimes necessary to express the problem clearly by adding two additional lines to your full description: *Expected Result* and *Actual Result*.

FIGURE 3.1 *The Legend of Zelda: Tears of the Kingdom*. One would reasonably expect Link's big metal spiked club not to poke through solid walls. Source: *The Legend of Zelda: Tears of the Kingdom*. © Nintendo

Expected Result describes the behavior that an average player would reasonably expect from the game if they followed the steps in the bug report. This assumption is based on the tester's knowledge of the game design documentation, the game's target audience, and precedents set (or broken) by other games, especially games in the same genre.

Actual Result describes the defective behavior. Here is an example:

```
1. Create a multiplayer game.

2. Click Game Settings.

3. Using your mouse, click any map on the map list. Remember the
   map you clicked on.

4. Press up or down directional keys on your keyboard.

5. Notice the highlight changes. Highlight any other map.

6. Click Back.

7. Click Start Game.

Expected Result: Game loads map you chose with the keyboard.

Actual Result: Game loads map you chose with the mouse.
```

Although the game loaded a map, it was not the map the tester chose with the keyboard (the last input device used). That is a bug, albeit a subtle one. Years of precedent creates an expectation in the player's mind that the computer will execute a command based on the last input it received. Because the map-choosing interface failed to conform to player expectation and general precedent, it could be confusing or annoying, so it should be written up in a bug report.

Use "Expected Result" and "Actual Result" steps sparingly. In many cases, defects are obvious. (See Figure 3.2.) Here is an example of "stating the obvious" in a crash bug.

```
4. Choose "Next" to continue.

Expected Result: You continue.

Actual Result: Game locks up. You must reboot the console.
```

It is understood by all members of the project team that the game should not crash. Do not waste time and space stating that with an unnecessary expected result statement.

You should use these statements sparingly in your bug reports, but you should use them when necessary. They can often make a difference when a developer wants to close a bug in the database by declaring it "by design," "working as intended," or "NAB" (Not a Bug).

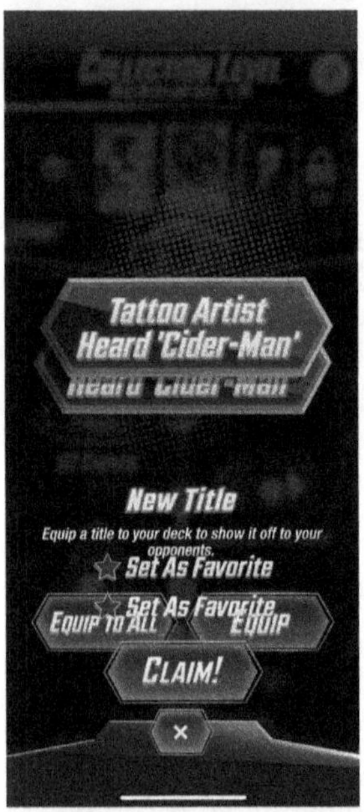

FIGURE 3.2 *Marvel Snap* (iOS): The double image here is obviously a bug. It is likely safe to skip the "expected result" statement in the bug report. Source: *Marvel Snap*. ©MARVEL ©Nuverse (Hong Kong) Limited ©Second Dinner

Habits to Avoid

For the sake of clarity, effective communication, and harmony among members of the project team, try to avoid two common bug-writing pitfalls: humor and jargon.

Although humor is often welcome in high-stress situations, it is not welcome in the bug database. There are too many chances for misinterpretation and confusion. During "crunch time," most people on the development team will be experiencing considerable stress. The defect database will be a significant source of that stress. Do not make the problem worse with attempts at humor (even if you think your joke is hilarious).

It perhaps seems counterintuitive to want to avoid jargon in such a specialized form of technical writing as a bug report, but it is wise to do so. Although some jargon is unavoidable, and each project team quickly develops its own nomenclature specific to the project they are working on, testers should avoid using (or misusing) too many obscure technical terms or in-game acronyms. Remember that your audience could range from programmers to financial or marketing executives, so use plain language as much as possible.

USING THE BUG DATABASE

Publishers and developers need an efficient way to keep track of the hundreds or thousands of bugs that can be uncovered by diligent game testers. The defect-tracking database (or "bug database," or sometimes just "bug base") is where each bug report is recorded, updated, and closed over the life of the project. A small student development team might only need a shared spreadsheet to keep track of the issues in their game. Larger projects require more robust database solutions. These can range from stand-alone database applications such as FileMaker Pro, or an open-source Web-based solution like Bugzilla, or a component of a project management suite such as Jira. For an online list of more defect tracking tool options, visit Software Testing Help "17 Best Bug Tracking Tools: Defect Tracking Tools Of 2024." (Software Testing Help 2023)

Although as a tester you will not have to be concerned about the installation and management of the bug database itself, you need to become familiar with how best to use it to record your defects and get them fixed, then closed.

Figure 3.3 shows the new defect entry window of the open-source MantisBT tool. The main elements of this window are the function selections along the sidebar, and the entry screen on the right. Field names with asterisks indicate that these fields are mandatory. You must provide information in each of these fields before your defect can be submitted.

In general, this function and the View Issues filters are very similar to using an email program. You can decide which things need your attention, browse through them, and make updates to issues when you need to.

The data entry screen is where you make the most of your contributions, so the following paragraphs explore some of the key fields you need to work with. To learn more about using other functions of MantisBT, you can explore the demo at *https://www.mantisbt.org/demo.php* (Mantis Bug Tracker 2024).

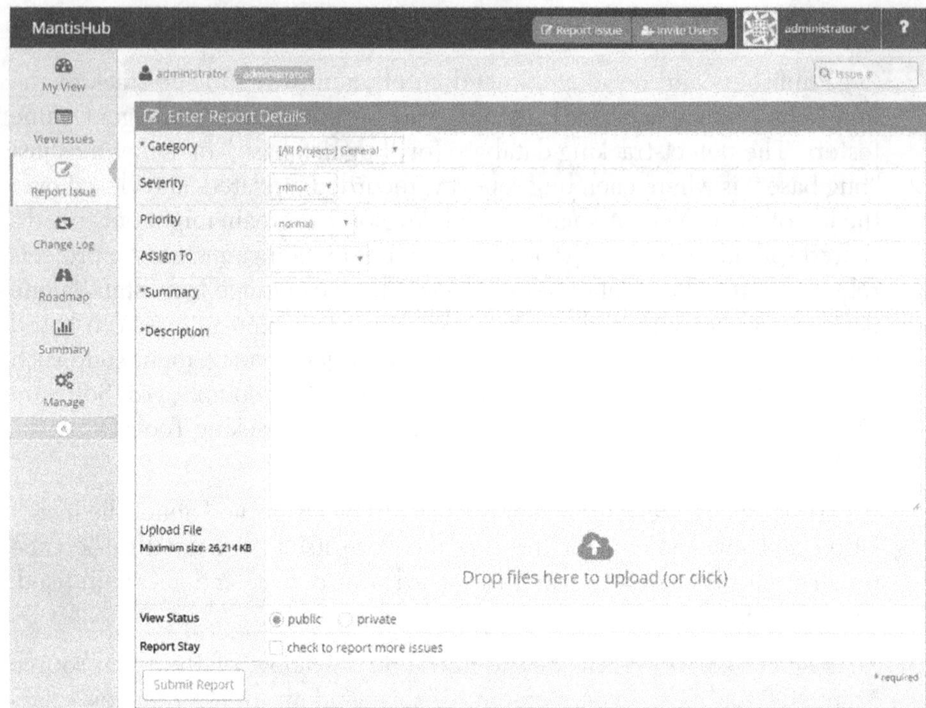

FIGURE 3.3 MantisBT "Report Issue" form

Category

Different users of the bug database will want to search and sort the reports for any number of reasons, so there are a number of fields, like *Category*, whose definitions and protocols will be determined based on the individual needs of your project. "Category" generally means "kind of bug," like crash or stop, network error, or even errors of art assets or animations.

Summary

This is where you type your brief description, or "headline," as discussed above. Remember to use the PAL (problem, action, location) rule to help you write a clear, specific summary.

(Full) Description

The *Description* field will hold as many characters as you need to write the steps necessary to reproduce the bug.

Severity

The *Severity* field is important for the routing and handling of your bug. The severity rating of a bug should be objective and based on criteria developed by the producer and lead tester during preproduction. Many studios use a rating system from A to D (or 1 to 4), based on the potential impact of the bug on the player. Severity A (or 1) bugs are fatal stops, crashes, freezes, or other catastrophic events. Severity B (or 2) bugs might be bugs that waste some of the players time, or break the immersion of the game. Severity C (or 3) bugs might be sloppy things like typos or grammar errors in text or texture rendering errors that do not impact gameplay, and so on. Figure 3.4 shows the defect Severity set to "major."

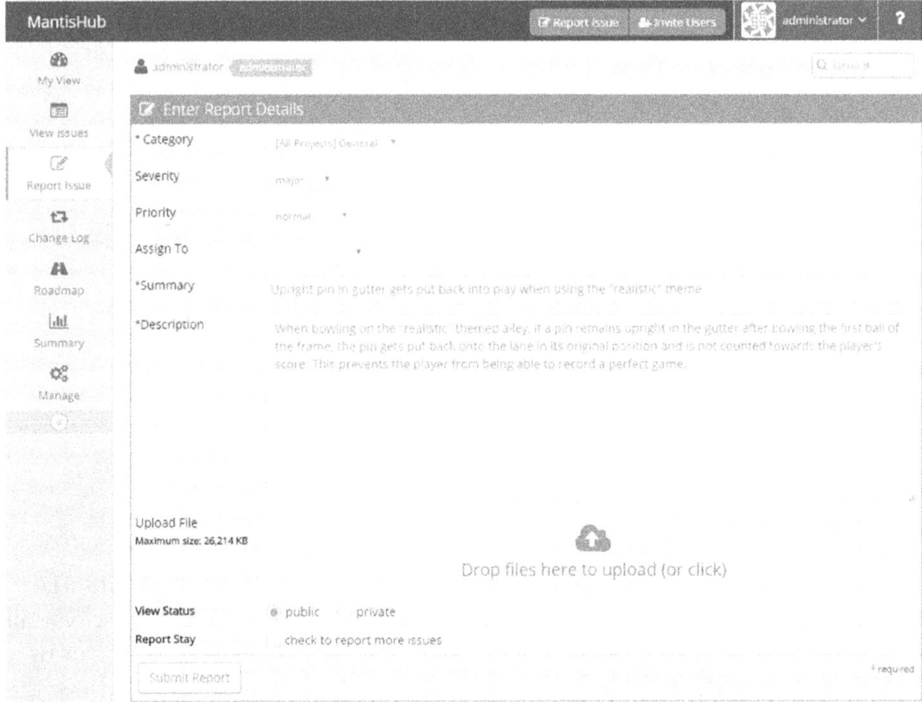

FIGURE 3.4 Severity Type selection

Priority

If severity is an objective judgment, *Priority* is subjective, and should be decided upon by a project manager based upon the day-to-day situation of the development team, where the team is in the project cycle, whether or not the affected feature is part of the current development focus, and other internal considerations. A beginning game tester should not be concerned with setting the priority of the bug, only in accurately rating its severity.

Attachments and Recordings

Finally, make sure you fill in any remaining required fields and include any other artifacts (like videos or annotated screenshots) or other information that might be of help to anyone trying to assess or fix the bug.

In MantisBT, you can use the Upload File function to add helpful files to the defect record. Attach or provide links to any of the following items that might be useful.

- Server logs
- Screenshots (with annotations and highlights, as needed)
- Sound files
- Save files
- A video recording (with sound) of the moments leading up to and including the appearance of the bug
- Traces of code in a debugger, if you have access to such resources
- Log files kept by the game platform, game engine, or hardware
- Operating system pop-ups and error codes
- Data captured by simulators for mobile development environments

Other Database Fields

Although these are not present in the MantisBT example interface, different projects may require additional fields in their databases. Not all defect tracking systems you use will have the same kind of structure or user interface as Mantis. Pay attention to getting the basics right and ask the other testers or your test lead what else is expected of you when reporting a bug. For example, you might be expected to send an email to a special mailing list if the defect tracking system you are using does not do that automatically, or you could be using a shared spreadsheet instead of a tool specifically designed for defect tracking. Some of these include the following:

Issue Number: This is the index field of the database. It is a unique number that identifies a specific bug report. It will be automatically generated as you create the report.

Status: This is the current state of the defect report. Values in this field might include new, open, closed, fixed, pending, verify fix, and NAB.

Frequency: This field indicates how often the bug will occur when the steps to reproduce are carefully followed. This is also known as *reproducibility rate*, or *repro rate*.

Author: This is the tester who first reported the bug.

Assigned to: This field indicates which team member is currently responsible for the bug, or who "owns" the bug. Producers may assign a bug to a particular programmer to be fixed. Once fixed, that programmer may assign the bug back to the author or QA lead so that the fix can be verified.

Metatags and Keywords are often added to a bug report to help someone find particular types of bugs, or all the bugs that affect a particular level, system, or character.

Interview: Jason Hutchins

"Quality Assurance is by nature adversarial," says industry veteran Jason Hutchins, who has worked as a senior producer on such franchises as *World of Warcraft*, *Madden*, and *Call of Duty*. He knows first-hand the role that game testing plays, because he rose up from the ranks of QA, beginning his career as a tester on *StarCraft* at Blizzard Entertainment.

QA is "there to say 'this is what we found, and what we found is a problem.' And games are so complex that there's *always* going to be problems." He notes that it is easy for developers to grow defensive about bugs and internalize the feedback from QA as criticism of their work. "If all I ever tell you is 'here's another mistake that I found'—with the implied *and you made it*—a tense relationship is going to happen."

He says it is very important everyone involved in the game development process to communicate clearly, professionally, and empathetically. "Testers need to be careful about how they phrase things, so as not to be insulting," he says. "We are all humans, and burdened with emotion."

"There are two things that you need in order to be good at QA, and they're at odds with each other," Mr. Hutchins says. The first is focusing on finding the fewest steps to a reproducible path for the bug. But he says the other is the ability to keep multiple hypotheses in your head as you try to find that path. "Because games are so complicated, there are many times when you can't focus on one hypothesis alone, because what happens if your test of that hypothesis fails?" he asks. "You then have to ask what other things are happening [in the game] that could have contributed to this."

Having worked with developers and testers for nearly three decades, Mr. Hutchins has seen the role of testing grow and become more appreciated, as developers learn the value of testing from the very beginning of a project. He is a proponent of developers using "embedded" testers working side-by-side with developers early in the process, in order to save development time. "You have much faster and much clearer communication on teams that share an office," he says.

He has three key pieces of advice for a novice game tester:

1. **Be clear** when writing the steps to recreate a bug, including both the expected result and the "bad" result.
2. **Be kind** to everyone on the project team, "whether or not they deserve it."
3. **Be curious** about how games are made. He suggests taking a programming course or following some basic tutorials in a game engine. "Game design is an art. It's not like developing banking software, where the systems design and user stories are more clearly defined."

(Permission: Jason Hutchins)

Although testing build after build might seem repetitive, each new build provides exciting new challenges with its own successes (fixed bugs and passed tests) and shortfalls (new bugs and failed tests). The purpose of going about the testing of each build in a structured manner is to reduce waste and to get the most out of both the game team and the test time. With each new build, you get new data that is used to refine test execution strategies and update or improve your test suites. From there, you prepare the test environment and perform a *smoke test* to ensure that the build is functioning well enough to deploy to the entire test team. Once the test team is working on the game, your top priority is typically regression testing to verify recent bug fixes. After that, you perform many other types of testing to find new bugs and to check that old ones have not re-emerged. New defects should be reported in a clear, concise, and professional manner after an appropriate amount of investigation. Once you complete this journey, you are rewarded with the opportunity to do it all over again.

EXERCISES

1. Briefly describe the difference between the Expected Result and the Actual Result in a bug report.
2. What is the purpose of regression testing?
3. The Brief Description field of a defect report should include as much information as possible. True or False?
4. A tester should write as many steps as possible when reporting a bug to ensure the bug can be reliably reproduced. True or False?
5. On a table in a kitchen is a loaf of bread, a jar of grape jelly, a jar of peanut butter, and a knife. Write step-by-step instructions to make a peanut butter and jelly sandwich. Assume the person reading the instructions has never seen, heard of, or eaten a peanut butter and jelly sandwich before.

REFERENCES

Mantis Bug Tracker. 2024. "Download | Setup Your Own." Accessed January 4, 2024. *https://mantisbt.org/demo.php*.

Software Testing Help. 2023. "17 Best Bug Tracking Tools: Defect Tracking Tools Of 2024." Last modified November 27, 2023. *https://www.softwaretesting-help.com/popular-bug-tracking-software*.

CHAPTER 4

HOW BUGS HAPPEN

Bugs happen.

Some programmers may claim that their code "doesn't have bugs," but they are wrong. No professional programmer wants their code to contain bugs, just as no professional writer wants their manuscript to contain typos or grammar errors. However, errors occur in game code for any number of reasons:

- It is common for game software to go wrong, since games often push the limits of the hardware they are played on.
- Game software is complex, so there are many opportunities to make a mistake.
- People write game software and people make mistakes.
- Software tools such as game engines are used to produce games, and these tools may themselves contain bugs.
- There is a great deal of money at stake for games to succeed, so pressure to perform quickly is felt by every member of the development team.
- Games must work on multiple platforms with a variety of configurations and devices.

- People expect more out of every new game.
- Games have to be fun, meet both business and player expectations, and get released on time.

A short statement that summarizes everything on this list is "games get made wrong." If you can identify mechanisms or patterns that describe how games get made wrong, you can relate that to what kinds of problems you should look out for and as you follow your path to becoming a skilled game tester. Maybe the people who care the most about game testing can help you to understand.

WHO CARES?

Testing must be important to game publishers because of all the trouble they go through to staff and fund testers and then to organize and manage the rounds of testing that precede the official game release. Testing is important to game console manufacturers because they require certain quality standards to be met before they will allow a title to ship for their platform. Mobile game testing is required by handset manufacturers, wireless carriers, and digital storefronts for them to approve games for sale on their devices, networks, and stores.

Testing is important to the development team. They rely on testers to find problems in the code. The testers sometimes bear the burden of getting blamed when serious defects escape their notice. If defects do escape, someone wonders why they paid all that money for testing.

Testing is important because of the contractual commitments and complex nature of the software required to deliver a top-notch game. Every time someone outside of your team or company sees the game, it is going to be scrutinized and publicized.

Despite all of the staffing, funding, and caring, games still get made wrong.

DEFECT TYPES

Software can fail in a variety of ways. It is useful to classify defects into categories that reveal how the defect was introduced and how it can be found or, even better, *avoided* in the future. The Orthogonal Defect

Classification (ODC) system, developed at IBM by Ram Chillarege, was created for this purpose (IEEE Computer Society n.d.). This system defines multiple categories of classification, depending on the development activity that is taking place. This chapter explores the eight Defect Type classifications and examines their relevance to game defects. The Defect Type classifies the way the defect was introduce into the code. Keep in mind that each defect can be the result of either incorrect implementation or code that is simply missing. The following ODC Defect Types summarize the different categories of software elements that go into producing the game code:

- Function
- Assignment
- Checking
- Timing
- Builds/Package/Merge
- Algorithm
- Documentation
- Interface

NOTE *For an expanded examination of ODC types and defect triggers, see Chapter 12, "Defect Triggers."*

Defect examples in this section are taken from the *Dark Age of Camelot* (DAOC) game Version 1.70i Release Notes, posted on July 1, 2004 (Fanbyte 2004). A predecessor to *World of Warcraft*, *Dark Age of Camelot* is also a Massively Multiplayer Online Role-Playing Game (MMORPG) that is continually updated to expand and enhance the player's game experience. As a result, it is patched frequently. This gives us the opportunity to examine it as it is being developed, as opposed to a game that has a single point of release to the public.

The defect description by itself does not tell us *how* the defect was introduced in the code—which is what the Defect Type classification describes. Because we do not have access to the development team's defect tracking system to know exactly how this bug occurred, let's take one specific bug and look at how it *could* have been caused, and therefore into which defect type it should be categorized.

Here is a fix released in the patch for *Dark Age of Camelot*, which is referenced above and throughout the examples in this chapter:

> "The Vanish realm ability now reports how many seconds of super stealth you have when used."

If that is how the ability is supposed to work, then you can imagine that the bug was reported with a description that went something like this:

```
The Vanish realm ability fails to report how many seconds
of super stealth you have when it is used.
```

The full details of the Vanish ability are as follows (Dark Age of Camelot 2024):

```
Provides the stealther with super stealth, which cannot
be broken. Also will purge DoTs and Bleeds and provides
immunity to crowd control. This ability lasts for 1 to
5 seconds depending on level of Vanish. The stealther
also receives an increase in movement speed as listed. A
stealther cannot attack for 30 seconds after using this
ability.

Effect:

L1 - Normal Speed, 3 second immunity

L2 - Speed 1, 3 second immunity

L3 - Speed 3, 4 second immunity

L4 - Speed 4, 5 second immunity

L3 - Speed 5, 6 second immunity

Type: Active

Re-use: 15 minutes.
```

Functions

A *function* error is one that affects a game capability or the way the user experiences the game. The code providing this function is missing or incorrect in some or all instances where it is required.

Here is an imaginary snippet that illustrates code which might be used to set up and initiate the Vanish ability. The player's Vanish ability level is passed to a handler routine specific to the Vanish ability. This routine is required to make all of the function calls necessary to activate this ability. The `g_vanishSpeed` and `g_vanishTime` arrays store values for each of the three levels of this ability, plus a value of 0 for level 0. These arrays are named with the "`g_`" prefix to indicate that they are global, because the same results apply for all characters that have this ability. Values appearing in all uppercase letters indicate that these are constants.

```
void HandleVanish(level)
{
    if (level == 0)
        Return;        //player does not have this ability so
leave
    PurgeEffects (damageOverTime);
    IncreaseSpeed(g_vanishSpeed[level]);
    SetAttack (SUSPEND, 30SECONDS);
    StartTimer(g_vanishTime[level]);
    Return;
}
```

Missing a call to a routine that displays the time of the effect is an example of a function type defect for this code. Maybe this block of code was copied from some other ability and the "vanish" globals were added but without the accompanying display code. Alternatively, there could have been a miscommunication about how this ability works and the programmer did not know that the timer should be displayed.

Alternatively, the function to show the duration to the user could have been included, but called with one or more incorrect values:

```
    ShowDuration(FALSE, g_vanishTime[level]);
```

TIP

No Coding Experience? No Problem!
You do NOT need to know how to read or write computer code to be an effective game tester. However, learning some coding skills along the way will be helpful to your career. Many people who hesitate to learn coding worry about the math involved, but coding lets you tell the computer to do the math for you! Think of learning more about coding as an opportunity to build your skills in logical thinking and precision, which are crucial for writing clear bug reports and understanding game mechanics like a pro.

Assignments

A defect is classified as an *assignment* type when it is the result of incorrectly setting or initializing a value used by the program or when a required value assignment is missing. Many of the assignments take place at the start of a game, a new level, or a game mode. Here are some examples for various game genres:

Sports

- Team schedule
- Initialize score for each game
- Initial team lineups
- Court, field, or rink where game is being played
- Weather conditions and time of day

Role-Playing Game (RPG), Adventure

- Starting location on map
- Starting attributes, skills, items, and abilities
- Initialize data for current map, including randomized placement of objects and enemies
- Initialize journal

Racing

- Initialize track/circuit data
- Initial amount of fuel or energy at start of race
- Placement of power-ups and obstacles
- Weather conditions and time of day

Casino Games, Collectible Card Games, Board Games

- Initial amount of points or money to start with
- Initial deal of cards or placement of pieces

- Initial ranking or seeding in tournaments
- Position at the game table and turn order

Fighting

- Initial health and energy
- Initial position in ring or arena
- Initial ranking or seeding in tournaments
- Arena where the fight will occur

Strategy

- Initial allocation of units
- Initial allocation of resources
- Starting location and placement of units and resources
- Goals for current scenario

First Person Shooter (FPS)

- Initial health, energy
- Starting equipment and ammunition
- Starting location of players
- Number and strength of CPU opponents

Puzzle Games

- Starting configuration of puzzle
- Time allocated and criteria to complete puzzle
- Puzzle piece or goal point values
- Speed at which puzzle proceeds

You can see from these lists that any changes could tilt the outcome in favor of the player or the CPU. Game designers pay careful attention to balancing all of the elements of the game. Initial value assignments are important for providing that game balance.

Even the Vanish effect could have been the result of an assignment problem. In the imaginary implementation that follows, the Vanish ability is activated by setting up a data structure and passing it to a generic ability handling routine.

```
ABILITY_STRUCT    realmAbility;
realmAbility.ability = VANISH_ABILITY;
realmAbility.purge = DAMAGE_OVER_TIME_PURGE;
realmAbility.level = g_currentCharacterLevel[VANISH_ABILITY];
realmAbility.speed = g_vanishSpeed[realmAbility.level];
realmAbility.attackDelay = 30SECONDS;
realmAbility.duration = g_vanishTime[realmAbility.level];
realmAbility.displayDuration = FALSE; // wrong flag value
HandleAbility(realmAbility);
```

Alternatively, the assignment of the `displayDuration` flag could be missing altogether. Again, cutting and pasting code from another part of the program could be how the bug was introduced, or it could have been wrong or left out as a mistake on the part of the programmer, or there could have been a misunderstanding about the requirements.

Checking

A *checking* defect type occurs when the code fails to validate data properly before it is used. This could be that a check for a condition is missing or improperly defined. Some examples of improper checks in C code would be the following:

- "=" instead of "==" used for comparison of two values
- Incorrect assumptions about operator precedence when a series of comparisons are not parenthesized
- "Off by one" comparisons, such as using "<=" instead of "<"
- A value `(*pointer)` compared to NULL instead of an address `(pointer)`—either directly from a stored variable or as a returned value from a function call
- Ignored (not checked) values returned by C library function calls, such as `strcpy`

Let's think about the Vanish bug again. The following shows a checking defect scenario where the ability handler does not check the flag for displaying the effect duration or checks the wrong flag to determine the effect duration:

```
HandleAbility    (ABILITY_STRUCT ability)
{
    PurgeEffect(ability.purge);
    if (ability.attackDelay > 0)
        StartAttackDelayTimer(ability.attackDelay);
    If (ability.immunityDuration == TRUE)
    //should be checking ability.displayImmunityDuration!
        DisplayAbilityDuration(ability.immunityDuration);
}
```

Timing

Timing defects have to do with the management of shared and real-time resources. Some processes could require time to start or finish, such as saving game information to a hard disk. Operations that depend on that data should not be prevented until completion of the dependent process. A common way of handling this is to present a transition such as an animated cut scene or a "splash" screen with a progress bar that shows the player that the information is being saved. Once the save operation is complete, the game resumes. Other timing-sensitive game operations include pre-loading audio and graphics so that they are immediately available when the game needs them. Many of these functions are now handled in the game hardware, but the software still might need to wait for some kind of notification, such as a flag that get set, an event that gets sent to an event handler, or a routine that gets called once the data is ready for use.

NOTE

The FMOD multi-platform audio engine illustrates how an audio event notification scheme is set up and utilized (FMOD 2024). To play a song, the developer starts by initializing FMOD, loading a song which returns a handle, and passing that handle to the PlaySong function. When an event is eventually detected that should stop the song, such as when the game environment changes to a new setting (city, arena, or planet), StopSong will do just what its name suggests and the handle can be freed using FreeSong.

User inputs can also require special timing considerations. Double-clicks or repeated presses of a button could cause special actions in the game. There could be mechanisms in the game platform operating system to handle this or the game team might put its own into the code.

In multiplayer games, data is constantly flowing between players and the game server(s). This information has to be reconciled and handled in the proper order or the game behavior will be incorrect. Sometimes the game software tries to predict and fill in what is going on while it is waiting for updated game information. When your character is running around, this can result in jittery movement or even a "rubber band" effect, where you see your character run a certain distance then, all of a sudden, you see them being attacked back where they started to run.

Getting back to the familiar Vanish bug, let's look at a timing defect scenario. In this case, pretend that one function starts up an animation for casting the Vanish ability, and a global variable `g_animationDone` is set when the animation has finished playing. Once `g_animationDone` is TRUE, the duration should be displayed. A timing defect can occur if the `ShowDuration` function is called without waiting for an indication that the Vanish animation has completed. The animation will overwrite anything that gets put on the screen. Here is what the defective portion of the code might look like:

```
StartAnimation(VANISH_ABILITY);
ShowDuration(TRUE, g_vanishImmunityTime[level]);
```

This would be the correct code:

```
StartAnimation(VANISH_ABILITY);
While(g_animationDone == FALSE)
    ; // wait for TRUE
ShowDuration(TRUE, g_vanishImmunityTime[level]);
```

Build/Package/Merge

Build/package/merge or, simply, *build* defects are the result of mistakes in using the game source code library system, managing changes to game files, or identifying and controlling which versions get built.

Building is the act of compiling and linking source code and game assets such as art assets, text, and sound files to create an executable game. Configuration management software is often used to help manage and control the use of the game files. Each file might contain more than one asset or code module. Each unique instance of a file is identified by a unique version identifier.

The specification of which versions of each file to build is done in a *configuration specification* (or *config spec*). Trying to specify the individual version of each file to build can be time consuming and error prone, so many configuration management systems provide the ability to label each version. A group of specific file versions can be identified by a single label in the config spec.

Table 4.1 shows some typical uses for labels. Your team perhaps will not use the exact label names shown here, but they will likely have similarly named labels that perform the same functions.

TABLE 4.1 Typical labels and uses

Label	Usage
[DevBuild]	Identifies files that programmers are using to try out new ideas or bug fix attempts.
[PcOnly]	Developing games for multiple platforms might require a different version of the same file that is built for only one of the supported platforms.
[TestRelease]	Identifies a particular set of files to use for a release to the testers. Implies that the programmer is somewhat certain the changes will work. If testing is successful, the next step might be to change the label to an "official" release number.
[Release1.1]	After successful building and testing, a release label can be used to "remember" which files were used. This is especially helpful if something breaks badly later on and the team needs to backtrack (or *roll back*) either to debug the new problem or to revert to a previous version.

Each file has a special evolutionary path called the *mainline*. A *version tree* provides a graphical view of all versions of a file and their relationship to one another with respect to the mainline. Figure 4.1 shows how a new version added to the mainline is represented on the version tree.

Any new versions of files that are derived from one already on the mainline are called *branches*. Files on branches can also have new branches

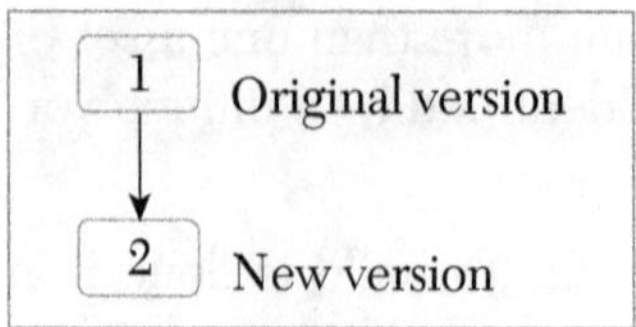

FIGURE 4.1 Mainline of a simple version tree

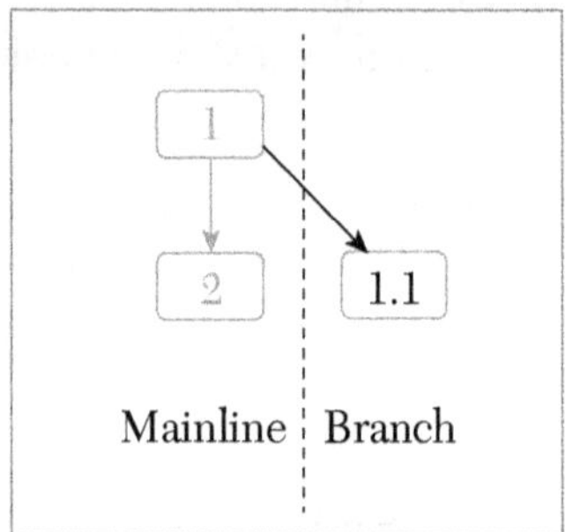

FIGURE 4.2 A version tree with a branch

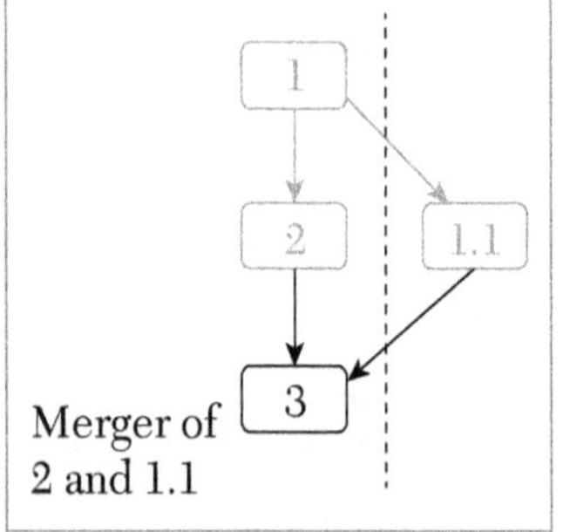

FIGURE 4.3 Merging back to the mainline

that evolve separately from the first branch. Figure 4.2 shows how a branch is numbered and represented graphically with respect to the mainline.

The changes made on one or more branches can be combined with other changes made in parallel by a process called a *merge*. Merging can be done manually, automatically, or with some assistance from the configuration management system, such as highlighting which specific lines of code differ between the two versions being merged together. See Figure 4.3 for an example of how a version tree can evolve as a result of branching and merging.

When a programmer wants to make a change to a file using a configuration management system, the file gets checked out. Then, once the programmer is satisfied with the changes and wants to return the new file as a new version of the original one, the file is *checked in*. If at some point in time the programmer changes his mind, the file check-in can be canceled and no changes are made to the original version of the file.

With that background, let's explore some of the ways a bug can be introduced during this complex and exacting process.

Specifying a wrong version or label in the configuration specification might still result in successfully generating a game executable, but it will not work as intended. It could be that only one file is wrong, and it has a feature used by only one type of character in one particular scenario. Mistakes like this keep game testers in business.

It is also possible that the configuration specification is correct, but one or more programmers did not properly label to the version that needed to be built. The label can be left off, left behind from an earlier version, or typed incorrectly so that it does not exactly match the label in the config spec.

Another problem can occur as a result of merging. If a common portion of code is changed in each version being merged, it will take skill to merge the files and to preserve the functionality in both changes. The complexity of the merge increases when one version of a file has deleted the portion of code that was updated by the version it is being merged with. If a person is doing the merges, these problems will perhaps be easier to spot than if the build computer is making these decisions and changes it entirely on its own.

Sometimes the code will give clues that something is wrong with the build. Comments in the code like `//TAKE THIS OUT BEFORE SHIPPING!` could be an indication that a programmer forgot to move a label or check a new version of the file back into the system before the build process started.

Referring back to Figure 4.3, assume the following for the Vanish code:

1. Versions 1 and 2 do not display the Vanish duration.
2. Version 1.1 introduced the duration display code.
3. Merging versions 2 and 1.1 produces version 3, but deletes the part of the code in version 1.1 that displays the duration (i.e., fixes the bug).

For the Vanish display bug, here are some possible build defect type scenarios:

- The merge that produced version 3 deleted the part of the code in version 1.1 that displays the duration. Version 3 gets built, but we get no duration display.
- Versions 1.1 and 2 were properly merged, so the code in version 3 will display the duration. The label used by build specification has not been moved up from version 2 to version 3, however, so version 2 gets built and we get no duration display.
- Versions 1.1 and 2 were properly merged, so the code in version 3 will display the duration. The build label was also moved up from version 2 to version 3. However, the build specification was hard-coded to build version 2 of this file instead of using the label, so we get no duration display.

Algorithms

Algorithm defects include efficiency or correctness problems that result from some calculation or decision process. Think of an algorithm as a process for arriving at a result (the answer is 42) or an outcome (the door opens). Each game is packed with algorithms that you might not even notice if they are working correctly. Improper algorithm design is often at the root of the ways people find to gain an unexpected advantage in a game. (These are often called *exploits*.) Here are some places where you can find algorithms and algorithm-type defects in games from various genres:

Sports

- CPU opponent play, formation, and substitution choices
- CPU trade decisions
- The individual AI behavior for all positions for both teams in the game.
- Determining camera angle changes as the action moves to various parts of the field
- Determining penalties and referee decisions
- Determining player injuries
- Player stat development during the course of the season
- Enabling special power-ups, awards, or modes

Role-Playing Game (RPG), Adventure

- Opposing and friendly character combat actions
- Damage calculations based on skills, armor, weapon type, and strength
- Saving throw calculations
- Determining the result of using a skill such as crafting or persuading
- Experience point calculations and bonuses
- Ability costs, duration, and effects
- Resource and conditions needed to acquire and use abilities and items
- Weapon and ability targeting, area of effect, and damage over time

Racing

- CPU driver characteristics, decisions, and behaviors (such as when to make a pit stop or use power-ups)
- Damage and wear calculations for cars, and damaged car behavior
- Rendering car damage
- Automatic shifting
- Factoring effects of environment such as track surface, banking, and weather
- CPU driver taunts

Casino Games, Collectible Card Games, Board Games

- Opposing player styles and degree of skill
- Applying the rules of the game
- House rules, such as when the dealer must stay in Blackjack
- Betting options and payouts
- Fair distribution of results; for example, no particular outcome (such as a card, dice roll, or roulette number) seems favored

Fighting

- CPU opponent strike (offensive) and block (defensive) selection
- CPU team selection and switching in and out during combat
- Damage/point calculation, including environmental effects
- Calculating and rendering combat effects on the environment
- Calculating and factoring fatigue

Strategy

- CPU opponent movement and combat decisions
- CPU unit creation and deployment decisions
- Unit pathfinding

- Damage and effect calculations
- Enabling the use of new units, weapons, technologies, and devices

First-Person Shooters

- Opposing and friendly character combat actions
- Damage calculations based on skills, armor, weapon type, and strength
- Environmental effects on speed, damage to player, deflection, or concentration of weapons

Puzzle

- Points, bonus activation, and calculations
- Determining criteria for completing a round or moving to the next level
- Determining success of puzzle goals, such as forming a special word, or matching a certain number of blocks
- Enabling special power-ups, awards, or modes

To complicate matters further, some game titles incorporate more than one genre, each with a different set of algorithms. For example, the *Pokémon SoulSilver* and *HeartGold* games mainly focus on following the storyline while training up your Pokémon to higher levels, but they also include beauty and athletic competitions, as well as a *Minesweeper*-style mini-game. The *Unreal Tournament* series is primarily considered a first-person shooter. It also incorporates adventure and sports elements at various stages of the tournament. Some sports games require players to acquire trading cards or other rewards using virtual currency earned from winning games and tournaments.

Some other areas where algorithm-type defects can appear in the game code are graphics rendering engines and routines, mesh overlay code, z-buffer ordering, collision detection, and attempts to minimize the processing steps to render new screens.

For the Vanish bug, consider an algorithm-defect scenario where the duration value is calculated rather than taken from an array or a file. Also suppose that a duration of 0 or less will not get displayed on the screen. If the calculation (algorithm) fails by always producing a 0 or negative number result, or the calculation is missing altogether, then the duration will not be displayed.

For the sake of an example, let's assume the immunity duration granted by Vanish were one second at Level 1, two seconds at Level 2, and five seconds at Level 3. This relationship can be expressed by the equation

```
VanishDuration = (2 << level) - level;
```

At Level 1, this becomes 2 - 1 = 1. For Level 2, 4 - 2 = 2, and at Level 3, 8 - 3 = 5. These are the results we want, according to the specification.

Now, what if by accident the modulus (%) operator was used instead of the left shift (<<) operator? This would give a result of 0 - 1 = -1 for Level 1, 0 - 2 = -2 for Level 2, and 2 - 5 = -3 for Level 3. The immunity duration would not get displayed, despite the good code that is in place to display this duration to the user. An algorithm defect has struck!

Documentation

Documentation defects occur in the fixed data assets that go into the game. This includes text, audio, and art assets, as listed here:

Text

- Dialogs
- User Interface elements (such as labels, warnings, and prompts)
- Help text
- Instructions
- Quest journals

Audio

- Sound effects
- Background music
- Dialog (human, alien, and animal)
- Ambient sounds (such as running water and birds chirping)
- Celebration songs

Video

- Cinematic introductions
- Cut scenes
- Environment objects
- Level definitions
- Body part and clothing choices
- Items (such as weapons and vehicles)

This special type of defect is not the result of faulty code. The errors themselves are in the bytes of data retrieved from files or defined as constants. This data is subsequently used by statements or function calls that draw art assets, print text on the screen, play audio, or write data to files. Defects of this type are detectable by reading the text, listening to the audio, checking the files, and paying careful attention to the art assets as they appear in the game.

String constants in the source code that get displayed or written to a file are also potential sources of documentation-type errors. When the game has options for multiple languages, putting string constants directly in the code can cause a defect. Even though it might be the proper string to display in one language, there will be no way to provide a translated version if the user selects an alternate language.

Bugs of this type are sometimes difficult to detect without specific knowledge of the content requirements of the game. For example, suppose you are making a game under license using the licensor's characters or other *intellectual property* (IP). This might include a game based on a popular movie franchise or animated program. You may have to test game content against the brand guide or other guidelines from the licensor to make certain that the colors, styles, logotypes, and general use of the licensed IP, to make certain that your game meets the standards of the licensor, and that they don't delay their approval of the game.

The examples in this section take a brief detour from the Vanish bug and examine some other bugs fixed in the *Dark Age of Camelot* 1.70i release, which appear at the end of the "New Things and Bug Fixes" list:

- `If something damages you with a DoT and then dies, you see "A now dead enemy hits you for X damage" instead of garbage.`

This could be a Documentation-type defect where a NULL string, or no string, was provided for this particular message, instead of the message text that is correctly displayed in the new release. There could be other causes in the code, however. Note that this problem has the condition "...and then dies," so maybe there is a checking step that had to be added to retrieve the special text string. A point to remember here is that the description of the defect is usually not sufficient to determine the specific ODC defect type, although it might help to narrow it down. Someone has to get into the bad code to determine how the defect got put in there.

- `Grammatical fixes made to bug report submissions messages, autotrain messages, and grave error messages.`

  This one is almost certainly a Documentation-type defect. No mention is made of any particular condition under which these are incorrect. The error is grammatical, so text was provided and displayed, but the text itself was faulty.

- `Sabotage ML delve no longer incorrectly refers to siege equipment.`

  This description refers to doing a */delve* command in the game for the Sabotage Master Level ability. The quick conclusion is that this was a documentation defect fixed by correcting the text. Another less likely possibility is that the delve text was retrieved for some other ability similar to Sabotage due to a faulty pointer array index—perhaps due to an assignment or function defect.

Interfaces

The last ODC defect type that needs to be discussed is the *interface* type. An interface occurs at any point where information is being transferred or exchanged. Inside the game code, interface defects occur when something is wrong in the way one module makes a call to another. If the parameters passed on do not match what the calling routine intended, then undesired results occur. Interface defects can be introduced in a variety of ways. Fortunately, these fall into logical categories:

1. Calling a function with the wrong value of one or more arguments
2. Calling a function with arguments passed in the wrong order
3. Calling a function with a missing argument
4. Calling a function with a negated parameter value

5. Calling a function with a bitwise inverted parameter value
6. Calling a function with an argument incremented from its intended value
7. Calling a function with an argument decremented from its intended value

Here is how each of these could be the cause of the Vanish problem. Let's use `ShowDuration`, which was introduced earlier in this chapter, and give it the following function prototype:

```
Void ShowDuration(BOOLEAN_T bShow, int duration);
```

This routine does not return any value, and takes a project-defined Boolean type to determine whether to show the value, plus a duration value, which is to be displayed if it is greater than 0. So, here are the interface-type defect examples for each of the seven causes:

1. `ShowDuration(TRUE, g_vanishSpeed[level]);`

   In this case, the wrong global array is used to get the duration (speed instead of duration). This could result in the display of the wrong value or no display at all if a 0 is passed.

2. `ShowDuration(g_vanishDuration[level], TRUE);`

   Let's say a `#define` statement causes the `BOOLEAN_T` data type to be an `int`, so inside `ShowDuration`, the duration value (first parameter) will be compared to TRUE, and the TRUE value (second parameter) will be used as the number to display. If the duration value does not match the `#define` for TRUE, then no value will be displayed. Also, if `#define` assigns TRUE a value of 0 or some negative number, then no value will be displayed because of our rule for `ShowDuration` that a duration less than or equal to zero does not get displayed.

3. `ShowDuration(TRUE);`

   No duration value is provided. If it defaults to 0 as a result of a local variable being declared within the `ShowDuration` routine, then no value will be displayed.

4. `ShowDuration(TRUE, g_vanishDuration[level] | 0x8000);`

   Here is a case where the code is unnecessarily fancy and causes trouble. An assumption was made that the high-order bit in the duration value acts as a flag that must be set to cause the value to be displayed. This could be left over from an older implementation of this function or a mistake made by trying to reuse code from some other function. Instead of the intended result, it changes the sign bit of the duration value and negates it. Because the value used inside of `ShowDuration` will be less than zero, it will not be displayed.

5. `ShowDuration(TRUE, g_vanishDuration[level] ^ TRUE);`

   More imaginary complexity here has led to an exclusive `OR` operation performed on the duration value. Once again, this is a possible attempt to use some particular bit in the duration value as an indicator for whether or not to display the value. In the case where TRUE is `0xFFFF`, this will invert all of the bits in the duration, causing it to be passed in as a negative number, thus altering its value and preventing it from being displayed.

6. `ShowDuration(FALSE, g_vanishDuration[level+1]);`

   This can happen when an incorrect assumption is made that the level value needs to be incremented to start with array element 1 for the first duration. When level is 3, this could result in a 0 duration, because `g_vanishDuration[4]` is not defined. That would prevent the value from being displayed.

7. `ShowDuration(FALSE, g_vanishDuration[level-1]);`

   Here, the wrong assumption is made that the level value needs to be decremented to start with array element 0 for the first duration. When level is 1, this could return a 0 value and prevent the value from being displayed.

Some of these examples may seem far-fetched, but they illustrate the variety of ways every single parameter of every single function call can be a ticking time bomb in the code. Once wrong move can cause a subtle, undetected, or severe interface defect.

Testing Happens

Any time someone plays a game, it is being tested. When someone finds a problem with the game, it makes an impression. A beta release is published for the express purpose of being tested. Hasn't the game already been extensively tested prior to the beta release? Why are problems still found by the beta testers? Even after the game is released to the general public, it is still being tested. Game companies scramble to get patches out to fix bugs. When they spend their time fixing old problems, that is time lost that could be spent on building new features and content, or the next game. Even patches can miss issues or create new problems that have to be fixed in yet another patch.

EXERCISES

1. Why is game testing important?
2. Which of the Defect Types do you think is the hardest for testers to find? Explain why.
3. List five situations where *assignments* are likely to occur in the code for a simulation game, such as games in *The Sims* or *Zoo Tycoon* series.
4. List five types of algorithms that you might find in a simulation game.
5. In the following code example from the publicly available source code for *Castle Wolfenstein: Enemy Territory*, identify line numbers (added in parentheses) that might be a source of a defect for each of the ODC Defect Types.

```
/*
===============
RespawnItem
===============
*/
(0)     void RespawnItem( gentity_t *ent ) {
(1)             // randomly select from teamed entities
(2)             if ( ent->team ) {
(3)                 gentity_t   *master;
(4)                 int count;
(5)                 int choice;
(6)                 if ( !ent->teammaster ) {
(7)                     G_Error( "RespawnItem: bad teammaster"
);
(8)                 }
(9)                 master = ent->teammaster;
(10)                 for ( count = 0, ent = master; ent; ent =
ent->teamchain, count++ )
(11)                     ;
(12)                 choice = rand() % count;
```

```
                for ( count = 0, ent = master; count < choice; ent = ent->teamchain, count++ )
                    ;
            }
            ent->r.contents = CONTENTS_TRIGGER;
            //ent->s.eFlags &= ~EF_NODRAW;
            ent->flags &= ~FL_NODRAW;
            ent->r.svFlags &= ~SVF_NOCLIENT;
            trap_LinkEntity( ent );
            // play the normal respawn sound only to nearby clients
            G_AddEvent( ent, EV_ITEM_RESPAWN, 0 );
            ent->nextthink = 0;
    }
```

6. That was fun! Let's do it again with another *Wolfenstein* example.

```
/*
============
G_SpawnItem

Sets the clipping size and plants the object on the floor.

Items can't be immediately dropped to floor, because they might
be on an entity that hasn't spawned yet.
============
*/
        void G_SpawnItem( gentity_t *ent, gitem_t *item ) {
            char    *noise;
            G_SpawnFloat( "random", "0", &ent->random );
            G_SpawnFloat( "wait", "0", &ent->wait );
            ent->item = item;
            // some movers spawn on the second frame, so delay item
            // spawns until the third frame so they can ride trains
            ent->nextthink = level.time + FRAMETIME * 2;
            ent->think = FinishSpawningItem;
            if ( G_SpawnString( "noise", 0, &noise ) ) {
             ent->noise_index = G_SoundIndex( noise );
        }
        ent->physicsBounce = 0.50;      // items are bouncy
```

```
        if ( ent->model ) {
        ent->s.modelindex2 = G_ModelIndex( ent->model );
        }
        if ( item->giType == IT_TEAM ) {
            G_SpawnInt( "count", "1", &ent->s.density );
            G_SpawnInt( "speedscale", "100", &ent-
>splashDamage );
            if ( !ent->splashDamage ) {
                ent->splashDamage = 100;
            }
        }
    }
```

NOTE *Both code excerpts above can be found at https://github.com/id-Software/Enemy-Territory/blob/master/src/game/g_items.c.*

REFERENCES

Dark Age of Camelot. 2024. "Realm Abilities." Retrieved January 8, 2024. *https://darkageofcamelot.com/content/realm-abilities.*

Fanbyte. 2004. "Can I get an I? 1.70i." Last Modified July 1, 2004. *https://camelot.allakhazam.com/story.html?story=3798.*

FMOD. 2024. "FMOD Studio: The adaptive audio solution for games." Retrieved January 5, 2024. *https://www.fmod.com/studio.*

IEEE Computer Society. n.d. "Ram Chillarege." Retrieved January 5, 2024. *https://www.computer.org/profiles/ram-chillarege.*

CHAPTER 5

THE PHASES OF GAME QUALITY ASSURANCE

Video games can range in size from tiny, downloadable, mobile phone games that take a few weeks to produce, to epic, massively multi-player online games developed over many years. No matter what size the game is or how long its production schedule, the testing of the game should always follow the same basic structure:

6. Pre-Production
7. Alpha
8. Beta
9. Gold
10. Post-Release

Like the plot of a suspense thriller, each sequence occurs more rapidly, and with much more heightened excitement—and stress—than the previous one. Figure 5.1 illustrates a rough timeline for a hypothetical mid-budget, hand-held racing game.

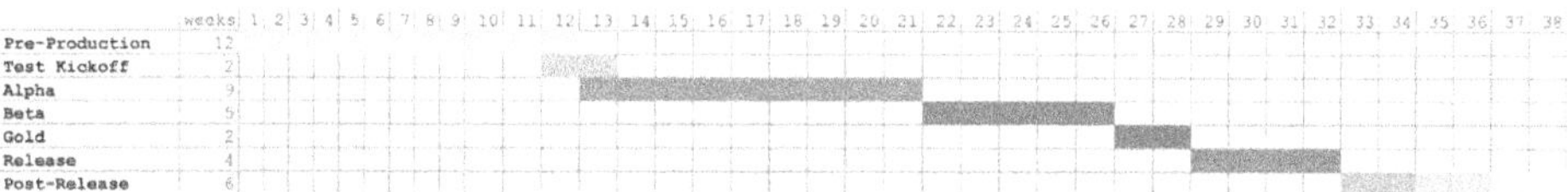

FIGURE 5.1 Hypothetical hand-held racing game timeline

If this timeline seems vaguely familiar, it may be because it reflects the game development life cycle we discussed in the first chapter. The following sections examine each phase in order to understand why it is vital to the project and distinct from the other phases.

PRE-PRODUCTION

Depending on your role on the team and when you were brought into the project, you might think that testing begins sometime after a good portion of the game is developed. In reality, testing begins when the project begins. There might not be people called "testers" involved at the beginning, but code, scripts, and assets are being produced from the start, and these need to be evaluated, critiqued, and corrected.

Much of what happens at the early stages of the project will determine how well testing will go later on. This includes how well the game does during the testing phase and how well the tests themselves are organized and executed. Both the QA team and the development team (or *dev team*) will have an easier job if more effort and skill are applied to testing activities at the beginning of the project. It is profoundly more difficult and expensive to try to make up for a lack of early testing by using more testers (and more overtime work) on the game in the later stages of development.

You cannot test quality into a game. The quality of the game is established by the code, the art, the audio, and the feedback loop between the game and the player—the "fun factor"—that is compiled into the software. All testing can do is tell the development team what is wrong with the software. Testing well and early can get problems fixed sooner and at a lower cost to the project.

If you received a coupon in the mail at the beginning of your project that said "Mail back this coupon to save 20% or more on your game budget," would you send it in? When you delay testing until the end of the project, it is the same as having that coupon but not mailing it in because you did not want to pay for the postage stamp.

Planning Tasks

Almost as soon as a game project is conceived, planning for its testing should begin. Test planning includes the tasks outlined in the following sections.

Determine the Scope of Testing the Project Will Require

The game design document (GDD), technical design document (TDD), and the project schedule are reviewed by the test manager to formulate a "scope of test" document that outlines the amount of testing resources (that is, the time, people, and money) they will need to get the game tested thoroughly for release.

The following section is a brief scope-of-test memo written by a publisher planning to develop an expansion pack to a real-time strategy (RTS) game released earlier that same year.

Expansion Plans

MEMORANDUM

To: Executive Producer

From: Manager of Quality Assurance

RE: RTS EXPANSION TEST PLAN SUMMARY

Summary

I have evaluated the GDD you forwarded last week. Assuming no changes to the game scope outlined in that document, it will take 1,760 hours to test the expansion pack, based on the following assumptions:

- 50-day production schedule
- four-person test team
- 10% allowance for overtime
- no post-release patch testing

Single Player (900 hours)

A significant amount of QA time will be spent testing the new campaign. Because the story mode of these missions will be highly script-dependent, testers will be tasked with breaking those scripts to ensure the user will have a seamless, immersive gameplay experience.

Because the developer has not designed cheats in the game, and because our experience with the original game was such that saved games could not reliably be brought forward from prior builds, campaign mode will take up the majority of test time.

Multiplayer (650 hours)

The thrust of multiplayer testing will be to

1. ensure correct implementation of new units and the new tile set
2. debug new maps
3. debug "interface streamlining" (the new functionality described in the design doc)
4. stress test game size
5. stress test army size
6. stress test game length
7. balance testing (as time permits)

Because the expansion pack introduces 12 new units, we will be concerned only with high-level balance testing. If one of the new units gives its clan an overwhelming advantage (or disadvantage), we would correct that defect. We do not have the resources available to re-evaluate each of the more than 50 existing units against the new units. We will count on the developers' design team (and user feedback compiled since the release of the original game) to fine tune the balance of the expansion pack.

Test Matrices (210 hours)

Because this a product for the PC and not a console, there will not be first-party TRC component to the testing. However, we will provide a similar standards-based level of final release testing based on a number of PC standards developed from our own experience and standards used at other PC publishers.

We will run the following standard matrices on the game:

1. install/uninstall matrix (with an emphasis on interoperability with the previous product)
2. Windows "gotchas" matrix
3. publisher standards matrix
4. multiplayer connectivity matrix

We will also produce and run the unit matrix developed while testing the original game on each new unit in the expansion pack.

Compatibility (0 hours)

Because the minimum system requirements will not change from the original game, we do not anticipate needing the services of a third-party hardware compatibility lab for compatibility testing. If any machine-specific bugs on the varied hardware in our internal lab appear during the normal course of testing, we will evaluate at that point whether a full compatibility sweep is warranted, along with an estimated additional budget.

Overtime (To Be Determined)

Because this product is projected by Marketing to have only a modest upside for the company, QA will work with Production to make best efforts to contain overtime costs. At this point, we anticipate working overtime only on such occasions that a failure to do so would make the product late.

Assign a Lead Tester

This is no trivial matter. The lead tester's experience, temperament, and skill set will have a tremendous influence over the conduct of the testing cycle. This might be the single most important decision the test manager makes on the project. A lead tester must be:

- **a leader** able to motivate the test team and keep them focused and productive
- **a team player** able to recognize the role testing plays as part of the larger production process
- **a communicator** able to gather and to present information clearly and concisely
- **a diplomat** able to manage conflicts as they arise

The test manager, or the lead tester, should then appoint an "assistant lead tester," often called a *primary tester*. On very large teams, it is not uncommon to have more than one primary tester, each leading specific sub-teams, such as multiplayer, story mode, and tutorials.

Establish Milestone Acceptance Criteria

It is advisable to require that any code release meets some criteria for being fit to test before you risk wasting your time, or your team's time, testing it. This is similar to the checklists that astronauts and pilots use to evaluate the fitness of their vehicle systems before attempting flight. Builds submitted to testing that do not meet the basic entry criteria are likely to waste the time of both testers and programmers. The countdown to testing should stop until the test "launch" criteria are met.

In an ideal world, you will be working from a contract, design specification, or product plan that defines very specific criteria for each phase of testing. However, the world is seldom ideal. The QA lead may need to work with the developer or publisher to clarify and specify the milestone definition criteria so that the test team can either "certify" a milestone candidate or reject it with feedback so that the developer knows what to include or fix so that the next milestone candidate can be accepted and the project can move forward.

The lead tester should take whatever materials are available and write a specification for the Alpha, Beta, and Gold (release) versions of the game. By establishing clear and unambiguous entry acceptance criteria for each phase of testing, conflicts can be avoided later in the project when pressure might be felt from various parts of the organization to begin, say, Beta testing on a build that is not truly a certified Beta version. Once the test manager has approved these criteria, they should be distributed to all senior members of the project team.

Three elements are required in the certification planning for each test phase:

1. **Entry Criteria:** The set of tests that a build must pass before entering a given test phase. The game will not be considered "at Alpha" until the code passes the Alpha Entry test, for example.

2. **Exit Criteria:** The set of tests that a build must pass before completing a test phase.

3. **Target Date:** The date both the development and test teams are working toward for a specific phase to launch.

Participate in Game Design Reviews

All stakeholders benefit from the QA team playing an active role from the beginning of the project. The lead tester or primary tester should participate regularly in design reviews. Their role is not to design the game, but rather to stay abreast of the latest design changes, as well as to advise the project manager of any technical challenges or testing complications that might arise from any anticipated feature revision. Changes in the scope of the game will dictate changes in the flow of the testing. The sooner that the tester leads are informed of a design change, the easier it is for them to adjust the test plan to accommodate the new design.

Set Up the Defect Tracking Database

This is a critical step, because a poorly designed database can waste precious minutes every time someone uses it, and those minutes quickly add up to hours or days toward the end of the project—time you will wish you had back! Figure 5.2 shows a typical entry in a bug database. (Note that the bug type "Unexpected Result" is too general.)

Battle Realms

Test Team ☐ Public Beta Test ☒ Publisher ☐ Developer

Bug # Crave 111 Bug Reporter Bob Bryant Date Entered 6/26/2001 Time Entered 4:58:08 PM

Bug Status In Progress Bug Severity B Bug Type: Unexpected Result Replication Rate: 100%

GameType LAN Game Version Version 2.03 Clan Serpent

Game Component Interface Map Level N/A Scenario N/A

Brief Description

Game Lobby: Serpent host changes to Wolf when Dragon player joins game.

Full Description

1. From Main Menu, Choose Multiplayer.
2. Choose LAN.
3. Click "Create Player."
4. Create a new player. Choose Serpent clan. Keep all other default settings.
5. Save this new player and click "Back."
6. Create a two-player game.
7. As Player 2 joins your game, watch your clan setting on the player list.

--> You change to Wolf clan as Player 2 joins game.

Assigned To Fix Type

Notes

FIGURE 5.2 Typical entry in a bug database

The lead tester and project manager should mutually agree on appropriate permissions, that is, those team members in each department who have editing rights to specific fields. The lead tester should also ask the project manager for a list of development team members to whom bugs will be assigned. The "assigned to" field allows the lead tester, project manager, or anyone else so entrusted to review new bugs and assign them to the right member of the development team. Programmers, artists, and other dev team members then search the database for bugs assigned to them and presumably fix their own defects. They can then assign the bug back to the lead tester so that the fix can be verified in the next build, and the bug can be closed.

Whether the bug database is going to exist on a secure internal server or be accessible over the Internet, it is a good idea at this point to populate the bug database with a few dummy records and double-check all passwords and permissions, both locally and remotely. Every person who will have access to the "bug base" should be assigned an individual password, and the lead tester can allow or block edit rights to individual fields based on the role that person will play on the project team.

Bug Database Setup Tips

A bug database that is editable by only the lead tester is not very useful; these tend to be static and incapable of conveying current information about the state of the project. Neither is a bug database in which every member of the team can edit every field; these are chaotic and ultimately useless.

In designing the bug database, the lead tester must balance the need for team members to communicate with each other about a particular defect with the equally important need to control the flow of information in order to manage task priorities. Programmers need to be able to comment on or to ask questions about a defect in the Developer Comments or Notes field, but they cannot be allowed to close a bug arbitrarily by changing the Status field to "closed" or "resolved." Testers need to be able to describe the bug in the Brief Description and Full Description fields, but they might not be qualified to judge who should "own" the bug in the Assigned To field.

Here are some recommendations:

- **Status** should be editable by the lead testers only. The default value for this field should be "New," so that as testers enter bugs, they can be reviewed and refined by the lead testers before they change the status to "Open" and the bug is then assigned to a member of the development team.
- **Severity** should be editable by the lead tester or primary testers. *Remember that the severity of a defect is not the same as its fix priority*. Testers, rightly, tend

to be passionate about the defects they find. It is the job of the test team leaders to check against this and assign severity in an objective manner.

- **Priority** should be editable by the project manager and senior members of the development team. This field is primarily a tool to help the project manager manage the flow of work to members of the development team. With the Agile development methodology becoming more popular in the games industry, project managers want maximum flexibility in assigning day-to-day or hour-to-hour priorities. Leave the priority field to them.
- **Category Fields** should be input by the testers and editable by the lead or primary testers. These fields may include such specifics as Game Type, Number of Players, Location, Bug Type, Reproduction Rate, and any other field that includes specific information about a bug that may need to be the basis of a search or sort of the database.
- **Brief and Full Descriptions** should be input by the testers and editable by the lead or primary tester. This is the heart of the bug report, including the steps to reproduce. It should not become a message board about the bug. Leave that to the Comments field.
- **Assigned To** is a field that should be editable by the lead tester and any member of the development team. The lead tester should typically assign new bugs to the project manager, who will then review the bug and assign it to a specific developer to be fixed. Once that bug is fixed, that person can either assign it back to the project manager for further review, or back to the lead tester so that the fix can be verified in the next build and the bug can be closed.
- **Developer Comments** should be editable by the project manager and any member of the development team.
- **QA Comments** should be editable by testers, the lead tester, the primary tester, and the QA manager.

Draft Test Plans and Design Tests

Having current and detailed knowledge of the game design is critical as the lead tester begins to plan for testing and draft test documents. An overall test plan document defines what types of tests will be done and what the individual test suites and matrices will look like (see the earlier chapter "Being a Video Game Tester"). This is the point in the project where you can put the methods described in the second half of this to good use. Remember: *Prior planning prevents poor performance.*

Test Plan

A *test plan* acts as the playbook for the QA team. It identifies the team's goals along with the resources (staff, time, tools, and equipment) and methods necessary to achieve them. Test goals are typically defined in terms of time and scope. The testing timeline often includes intermediate goals for one or more milestones that occur prior to the final release of the game. Any risks that could affect the test team's ability to meet the test goals are identified in the test plan, along with information about how to manage those risks if they occur. The scope of a test plan can be limited to a single subsystem of the game, or it can span many game features and releases. If the game is being tested at multiple sites, the test plan helps to identify what test responsibilities are assigned to each team. Appendix C contains a basic outline for a test plan.

Test Case

A *test case* describes an individual test that is to be performed by a tester or testers. Each test case also describes what operations to perform in order to meet its objective. Each individual operation within a test case is a *test step*. The level of detail in the test case can vary based on the standards of a particular test organization. Test cases are conceived and documented by each tester who is assigned a set of responsibilities in the test plan. The total set of test cases produced by a tester should fully cover their assigned responsibilities.

Test Suite

A *test suite* is a collection of related test cases that are described in specific detail. The test suite gives step-by-step instructions about what operations to perform on the game and what details to check for as a result of each step. These instructions should be sufficient for manual execution of the test or for writing code to automate the test. Depending upon how the test cases are written, they might or might not depend on the steps that were taken in a previous test case. Ideally, each test in the suite can be individually identified and executed independently of the other tests in the suite. Think of the test cases as individual chapters, while the test suite is a book that puts the test cases together into a detailed, cohesive story.

Testing Before Testing Begins

You might soon begin to get proto-builds in bits and pieces, with requests from the dev team to do very narrowly focused testing of certain specific features to give them confidence that these bits of code are working as intended before writing more code alongside them. This is sometimes called *modular testing*, because you are testing individual "modules" or features, not a complete build of the game.

At this stage in development, it is entirely likely that as code becomes functional and as modules are tested, the design of the game might be revised significantly "on the fly." Patience is required as you revise (and revise again) your test documents accordingly. Just as game design is an iterative process, game test materials should iterate as well.

During modular testing, it is premature to begin writing bugs beyond the narrow scope of the module's test case. True bug testing of the game will not begin until the dev team submits the first Alpha candidate.

Finally, the lead tester should begin to recruit or hire additional team members as necessary, according to the resource plan. Once the core test team is in place and the developers start to submit "pre-Alpha" builds of the game, then those builds can be tested against the milestone entry criteria.

Test Kickoffs

Kickoffs have a positive impact on game development, leading to better process definition, better problem solving, and schedule reduction. On a team in which testers have various levels of testing and game project experience, individual needs are not likely to be addressed at the project kickoff. Rather, it benefits the team to have kickoffs at the next-lowest level: a test kickoff for each "test" that is being created or executed by individual testers. The test kickoff illustrates the principle that increasing an organization's speed results from an iterative process of identifying obstacles, designing new processes that eliminate them, and ensuring that the new method is implemented. Just as in Agile development, test kickoffs can help the test team check in with each other at key milestones in the schedule to evaluate and improve their test tools and methods. Especially in the post-COVID era in which many testers may be working from home either full- or part-time, conducting test kickoffs is a good habit to ensure that all team members are and remain "on the same page."

Test kickoff activities are broken into two parts: tester preparation and the kickoff meeting, which is conducted according to the kickoff agenda. The tester's preparation steps and the kickoff agenda are documented in a test kickoff checklist, as shown in Figure 5.3.

TEST KICKOFF CHECKLIST

version 01.00

Game/Feature ______________________________

Tester ____________________ **Date** ______________

Tester Preparation

- ☐ Read the requirements for the feature being tested
- ☐ Gather equipment needed per the test equipment list
  - game platform(s) hardware
  - monitor
  - cables
  - controllers
  - save files
  - updates/patches/mods
  - test instruments
- ☐ Read the test script/report

Kickoff Agenda

Kickoff Leader

- Gives feature overview
- Addresses feature questions
- Brings up special instructions
- Brings up and solicits relevant improvement suggestions
- Addresses test execution questions/issues

FIGURE 5.3 Test kickoff checklist

From the test kickoff checklist, the tester should prepare in the following ways:

1. Read the requirements and documentation for the game feature being tested.
2. Gather equipment, files, and programs needed for test.
3. Read through the tests, making certain everything is clear and able to be performed.

The tester should consult with a "test expert" if there seem to be any roadblocks or questions regarding the completion of any preparation activities. The test expert can be the original author of the test, a tester who already has a lot of experience with the game feature, or the test lead. The expert should also be familiar with the recent defect history of the game or feature(s) to be tested. Experienced testers should not be exempt from this preparation process—just as "familiarity breeds contempt," overconfidence breeds carelessness, and this process should be completed fully before conducting the kickoff meeting.

Once the tester has completed the preparation activities, the test lead conducts the kickoff meeting by doing the following:

1. giving a feature overview
2. addressing feature questions
3. bringing up any special test instructions
4. bringing up and soliciting any relevant test improvement suggestions
5. addressing any test execution questions or issues
6. recording important issues on the kickoff form and providing a copy to the tester after the meeting is completed

Following the preparation steps listed on the checklist and participating in the meeting, per the kickoff agenda, benefits testing in the following ways:

- **prepares and equips** the tester to run through the entire test without stopping for equipment or questions

- **familiarizes** the tester with the *expected behavior* of the game or module during testing to increase tester awareness of "right" from "wrong"
- **resolves any conflicts** in test instruction prior to executing the test in order to eliminate retesting because of test ambiguities or errors
- **provides a forum** for test improvement at the grassroots level, improving tester involvement in and ownership of the test process

Each test kickoff is an opportunity to improve test understanding, test quality, and test execution. These opportunities would have been missed or identified much later in the test phase if the kickoff process was not used. The net result is that the test kickoff acts as a "pre-mortem" which identifies important issues prior to performing the test, rather than waiting to identify them in a postmortem evaluation after testing has already been done. As kickoff records are collected, systemic issues can be identified and addressed in the *current* test phase. Checklists, group meetings, and email are all means of communicating the lessons learned from the kickoffs and suggesting remedies to implement on the *current* project, rather than the next project.

By collecting and evaluating the results of kickoffs for each project, actions can be taken to prevent repeating any problems in future test efforts. The careful analysis of test kickoff results and the time savings achieved by using kickoffs can improve the way hundreds of other tests will be conducted going forward. The across-the-board use of test kickoffs will translate into further improvements in the test schedule and uncover more defects, leading to better game quality.

The following behaviors, which are driven by the use of test kickoffs, can reduce the length of the testing critical path:

- **Make fewer mistakes:** The test kickoff steps are designed to ensure that testing does not begin until the tester is fully equipped to test and understands the details and goals of the specific test. Among other things, this results in quicker and more accurate measurement of results.
- **Wasting less time:** As part of preparation, the tester reviews the test and requirements in their entirety. This reduces misunderstood and

improperly performed steps, resulting in much less test effort spent on backing up and redoing test sections.

- **All effort results in something that will be used:** Metrics show that the use of test kickoffs *reduces* the testing cycle time, even when you add in the time it takes to plan and hold the kickoffs.
- **Truth-telling is encouraged:** The one-on-one setting of a test kickoff is less intimidating than the group setting of a phase or release kickoff. The kickoff leader should make the tester comfortable and remind the tester of the kickoff goals. When testers see that their feedback results in improvements, they are more open about voicing their opinions and ideas.
- **Produce constructive discussions rather than destructive debates:** The test kickoff meeting gets every tester involved in process improvement. It also gives the tester and kickoff leader shared responsibility to address the issues raised and recorded in the meeting. Sticking to the kickoff agenda keeps the meeting focused on test-related issues.

NOTE *The idea that having a meeting would actually save time may seem counterintuitive to most people. We have held kickoffs for tests side-by-side with testing conducted without kickoffs. Our metrics show that the "kicked-off" tests were executed at 1.4 times the rate of the "non-kicked-off" tests. Putting it another way, testers who benefited from a kickoff meeting completed 40% more tests than those who did not have a kickoff.*

Test kickoffs can provide the same benefits for test creation as they can for test execution. Whether it is test flow diagrams (TFDs) and combinatorial tables (both discussed elsewhere in the book), test trees, matrices, or checklists, the process of creating test tools is made more efficient by using a kickoff process. All it takes is a slightly different agenda and checklist, as shown in Figure 5.4.

GAME TEST CREATION KICKOFF CHECKLIST

version 01.00

Game/Feature ______________________________

Tester ____________________ **Date** ______________

Test Creator Preparation

- ☐ Read the requirements for the feature being tested
- ☐ Read existing test scripts from similar features and/or games

Kickoff Agenda

Kickoff Leader

- Gives feature overview
- Addresses feature questions
- Brings up special instructions
- Brings up and solicits relevant improvement suggestions
- Addresses test case questions/issues

FIGURE 5.4 Test creation kickoff checklist

Version Control: Not Just for Developers

A fundamental principle of software development is that every build of an application should be tested as a separate and discrete version. Inadvertent blending of old code with new is one of the most common (and most preventable) causes of software defects. The process of tracking builds and ensuring that all members of a development team are checking current code and assets into the current version is known as *version control*.

Test teams must practice their own form of version control. There are few things more time-wasting than for a test team to report a great number of bugs in an old build. This is not only a waste of time, but it can cause panic on the part of the programmers and the project manager.

Proper version control for the test team includes the following steps:

1. Collect all prior physical (disk- or cartridge-based) builds from the test team before distributing the new build. The prior versions should be stored together and archived until the project is complete. (When testing digital downloads, uninstall and delete or archive prior digital builds.)
2. Archive all paperwork. This includes not only any build notes you received from the development team, but also any completed test suites, screen shots, saved games, notes, video files, and any other material generated during the course of testing a build. It is sometimes important to retrace steps along the paper trail, whether to assist in isolating a new defect or determining in what version an old bug was re-introduced.
3. Verify the build number with the developer prior to distributing it.
4. In cases where builds are transmitted electronically, verify the byte count, file dates, and directory structure before building it. It is vital in situations where builds are sent via secure FTP, email, a cloud-based file sharing solution, or other digital means, that the test team makes certain to test a version identical to the version the developers uploaded. Confirm the integrity of the transmitted build before distributing it to the testers.
5. Renumber all test suites and any other build-specific paperwork or electronic forms with the current version number.
6. Distribute the new build for smoke testing.

Configuration Preparation

Before the test team can work with the new build, careful preparation should be made. The test equipment must be readied for a new round of testing. The test lead must communicate the appropriate hardware configuration to each tester for this build. Configurations typically change little over the course of game testing. To test a single-player-only console game, you need the game console, a controller, and a memory card or hard drive. That hardware configuration will not change for the life of the project. If, however, the new build is the first in which network play is enabled, or a new input device or PC video card has been supported, you will perhaps need to augment the hardware configuration to perform tests on that new code.

Perhaps the most important step in this preparation is to eliminate any trace of the prior build from the hardware. "Wiping" the old build of a cartridge based-game on a Nintendo Switch™ is simple, because the only recordable media for that system is a microSD™ card or its internal flash memory drive. All you have to do is remove and archive the save game you created with the old build. More careful test leads will ask their testers to go the extra step of reformatting the media, which completely erases it, to ensure that not a trace of the old build's data will carry forward during the testing of the new build.

Save Your Saves!
Always archive your old player-created data, including game saves, options files, and custom characters, levels, or scenarios.

Not surprisingly, configuration preparation can be much more complicated for PC games. The cleanest possible testing configuration for a PC game is:

- a fresh installation of the latest version of the operating system, including any patches or security updates.
- the latest drivers for all components of the computer. This not only includes the obvious video card and sound card drivers, but also game controller drivers, chipset drivers, motherboard drivers, Ethernet card drivers, and even Wi-Fi firmware.

- the latest version of any "helper apps" or middleware the game requires to run. These can range from DirectX libraries to third-party launcher apps.

The only other software installed on the computer should be the new build.

Chasing False Bugs

We once walked into a QA lab that was testing what was at the time a cutting-edge 3D PC game. Testing of the game has fallen behind, and we had been sent from the publisher to investigate. We arrived just as the lab was breaking for lunch and were appalled to see the testers exit the game they were testing and fire up email, instant messenger clients, Web browsers, and file sharing programs—a host of applications that were installed on their test computers. Some even jumped into a game of *Unreal Tournament*. We asked the assistant test manager why he thought it was a good idea for the testers to run these extraneous programs on their testing hardware. "It simulates real-world conditions," he shrugged, annoyed by our question.

As you may have already guessed, this lab's failure to wipe their test computers clean before each build led to a great deal of wasted time chasing false defects—symptoms that testers thought were defects in the game, but which were in fact problems brought about by other programs running the background, taxing the computer's resources and network bandwidth. This wasted test time also resulted in much wasted programming time, as the development team tried to figure out why the game code might be causing such (false) defects.

The problem was solved by reformatting each test PC, freshly installing the operating system and latest hardware drivers, and then using a drive image backup program to create a system restore file. From that point forward, testers merely had to reformat their hard drive and copy the system restore file over from a CD-ROM in order to make their machines ready to test the next build.

Testing takes place in a "lab," and labs should be clean. So should test hardware. It is difficult to be too fastidious or concerned when preparing test configurations. When you get a new build, reformat your PC rather than merely uninstall the old build.

Clean Between Builds!
Delete your old builds. Reformat your test hardware (PCs, tablets, or smartphones). If it is a browser game, delete the browser cache.

Whatever protocol is established, *config prep* is crucial prior to the distribution and testing of a new build.

Smoke Testing

The next step after accepting a new build and preparing to test it is to certify that the build is worthwhile to submit to formal testing. This process is sometimes called *smoke testing* because it is used to determine whether a build "smokes" (malfunctions) when run. At a minimum, it should consist of a *load and launch*, that is, the lead or primary tester should launch the game, enter each module from the main menu, and spend a minute or two playing each module. If the game launches with no obvious problems, it is safe to log the build and duplicate it for distribution to the test team.

ALPHA TESTING

Now it is time to get busy. The project manager delivers you an Alpha candidate. You certify it against the Alpha milestone criteria you established in the planning phase. Full-bore testing can begin at last.

Over the course of Alpha testing, the game design may undergo significant revisions. Features are play tested and tweaked (or thrown away). Missing assets are integrated. Systems developed by different programmers are linked together. It is an exciting time.

As each member of the code and art team checks new work into the build, they are also (inadvertently) checking for new defects. This means that the game at this phase is a "target-rich environment" for a tester. It can also seem very overwhelming. It is critical at this stage that the test suites are strictly adhered to but adapted as necessary based on changes in the game's scope and design. Test suites will provide a structure for bringing order to what may seem like chaos.

Over the course of Alpha testing, all modules of the game should be tested at least once, and performance baselines should be established (frame rate, load times, and so on). These baselines will help the development team to determine how far they have to go to get each performance standard up to the target for release. For example, a frame rate of 30 (or even 15) frames of video per second (fps) might be acceptable in the early stages of developing a 3D action game, but the release target might be a solid 60 fps with no prolonged dips when there are a greater-than-usual number of animations and effects on the screen.

Alpha Phase Entry Criteria

The following are examples of possible Alpha entry criteria for a typical console game.

1. **All major game features exist and can be tested.** Some might still be in separate modules for testing purposes.
2. **A tester can navigate the game along some path from start to finish.** This assumes the game is linear, or has some linear component (for example, career or story mode in a sports game). Because many games are non-linear, the lead tester and project manager must agree ahead of time on a content completion target for such games (for example, six of 18 mini-games).
3. **The code passes at least 50% of the platform Technical Requirements Checklist.** Each console game has a set of standards published and tested against by the manufacturer of that platform. When you produce a PlayStation® 5 game, for example, the Global Format QA team at Sony Interactive Entertainment (SIE) will test it against the current PS5 technical requirements checklist (TRC) to make certain that the game complies with platform conventions. These requirements are very exacting, such as specifying the precise working of status or error messages a game must display during the same process.
4. **Basic user interface is complete and preliminary documentation is available to QA.** The main menu, most submenus, and the in-game interface (sometimes called the Heads-Up Display, or HUD) should be functional, if not yet finalized and visually polished. *Preliminary documentation* in this context means any explanation of new functionality, changed controller maps, and cheat codes (if any).

5. **The game is compatible with most specified hardware and software configurations.** For a cross-platform game, this means that the game will run on every targeted platform slated for initial commercial release. For a PC game, this criterion dictates that the game must run on a variety of computer systems with different hardware configurations (a range of CPU speeds, different video cards, RAM caches, and so on) that are still above the minimum system requirements as established in the TDD.

6. **Level scripting is implemented.** This pertains primarily to single-player story mode, or to the main quest line in an open-world game or RPG. An Alpha candidate that requires the tester to load separate levels manually would fail this criterion.

7. **First-party controllers and other peripherals work with the game.** Each platform manufacturer either makes or licenses for manufacture its own line of peripherals. Because support of these first-party peripherals is required by the platform TRC, and because the majority of testing will be done using first-party peripherals, they need to be supported in the Alpha build.

8. **Final or placeholder art is implemented for all areas of the game.** All the levels and characters must be textured and animated, though these textures, animations, and even the level geometry, might be subject to refinement or replacement as the game approaches Beta.

9. **Online multiplayer can be tested.** Enough networking code must be implemented so that at least two consoles can connect to the game sever and play a game.

10. **Placeholder audio is implemented.** It is entirely possible that the voice recording and final mixing sessions have not yet taken place at Alpha. In this case, members of the development team might record placeholder dialog and sound effects and integrate them where needed.

Regression Testing

Once a new build is distributed, it is *almost* time to test for new bugs. But before testing can take a step forward, it must first take a step backward and verify that the bugs which the development team claims to have fixed in this new build are indeed fixed. This process is known as *regression testing*.

Fix verification can be both very satisfying and very frustrating. It gives the test team a good sense of accomplishment to see the defects they report disappear one by one. It can be frustrating, however, when a fix of one defect creates another defect elsewhere in the game, as can sometimes happen.

The test suite for regression testing is the list of bugs the development team claims to have fixed. This *fix list*, sometimes called a *knockdown list*, is ideally communicated through the bug database. When the programmer or artist fixes the defect, all they have to do is change the value of the Developer Status field to "Fixed." This allows the project manager to track the progress on a minute-to-minute basis. It allows the lead testers to sort the regression set of "Fixed" bugs further for distribution to the test team (by bug author or by level, for example). At a minimum, the knockdown list should take the form of a list of bug numbers sent from the development team to the lead tester.

Always Get a Fix List!
Do not accept a build into test unless it is accompanied by a knockdown list. It is a waste of the test team's time to regress every open bug in the database every time a new build enters test.

Each tester will take the bugs they have been assigned and perform the steps in the bug report to verify that the defect is fixed. If they cannot reproduce the bug in the new build, then the bug has likely been fixed. The fixes for many defects are easily verified (such as typos and missing features). Some defects, such as hard-to-reproduce crashes, could *seem* fixed, but the lead tester should err on the side of caution before closing the bug by changing its status to *closed*. By flagging the defect as *verify fix*, the bug can remain in the regression set (stay on the knockdown list) for the next build (or two), but out of the set of open bugs that the development team is still working on. Once the bug has been verified as fixed in two or three builds, the lead tester can then close the bug with more confidence.

At the end of regression testing, the lead tester and project manager can get a very good sense of how the project is progressing. A high fix rate (number of bugs closed divided by the number of bugs claimed to have been fixed) means the developers are working efficiently. A low fix rate could be cause for concern. Are the programmers arbitrarily marking bugs as fixed if they think they have implemented new code that might address the defect, rather than troubleshooting the defect itself? Are the testers not writing clear bugs? Is there a version control problem? Are the test systems configured properly? While the lead tester and project manager mull over these questions, it is time for you to move on from regression to the next step in testing the new build: structured tests and exploratory tests.

BETA TESTING

By the end of Alpha, the development team should have a clear idea of the game they are creating. The development team has, for the most part, stopped creating new code and new artwork, and will now shift their focus to perfecting what they have already created. It is time to identify and fix the remaining bugs.

Although the term *Beta testing* frequently refers to any outside testing, it is only at the early stages of the Beta phase that final game play testing should take place with people outside the design team. The majority of testing done by outside Beta testers during true Beta is bug reporting and load testing. Gameplay feedback and suggestions should continue to be recorded for possible post-release implementation in a patch or sequel.

Beta Phase Entry Criteria

The following are examples of possible Alpha entry criteria for a typical console game:

1. **All features and options are implemented.** The game is "feature complete."
2. **The code passes at least 100% of the platform TRC.** Toward the end of Beta, the game should be ready for "pre-certification" submission to the platform manufacturer, if they offer such an opportunity. This process allows the platform manufacturer's QA team to test the game against the latest TRC and warn of any potential compliance issues.
3. **The game can be navigated on all paths.** Any bugs that might have closed off portions of the game are eliminated.
4. **The entire user interface is final.**
5. **The game is compatible with all specified hardware and software configurations.**

6. **The game logic and AI are final.** Programming is complete. The game knows its own rules. All AI or "bot" profiles are complete.
7. **All controllers work.** Those third-party peripherals that have been chosen by the development team (and the publisher) to be supported function with the game.
8. **Final artwork is implemented.** There should be no placeholder art left in the game. Beta is the phase when most screenshots, trailers, and gameplay footage will be taken to use to market the game.
9. **Final audio is implemented.** All placeholder audio has been replaced with final assets of the voice talent. (There might be a few do-over, or "pick-up," lines that have yet to be integrated, but these should not have an impact on in-game event timing or level scripting.)
10. **All online modes are complete and testable.**
11. **All language version text is implemented and ready for simultaneous release.** The game script (both written and spoken) and all on-screen text is locked and can be sent forward for translation and integration into the foreign-language versions of the game.

Design Lock

At some point during Beta testing, the project manager should declare the game to be in a state of *design lock* (sometimes called *feature lock*). The playtesting has concluded. Questions of balance have been resolved as best they can. The focus of the test team at this point should be to continue to run the test cases against the builds in an iterative manner, because each defect fixed at this point might have introduced another defect elsewhere in the game.

Toward the end of Beta, many tough decisions must be made. The teams are tired, tempers are on edge, and time is running out. In this charged atmosphere, the project team leaders have to make such critical choices as:

- **whether to implement that last-minute feature enhancement.** The designers might have had a great idea at the eleventh hour and are eager to introduce a new feature, character, or level. The project team leaders must weigh the risks of implementing the new feature (and possibly introducing new bugs and schedule slippage) against shipping a perhaps less compelling game on time.

- **whether to cut that level that just does not seem to be much fun.** Occasionally, it becomes clear during the course of testing that a level or other content component is a "problem child," and requires too much work to polish relative to the time left in the schedule. Cutting it out entirely could be problematic, however, in that the game will require new tests to ensure that the remaining levels run seamlessly around the deleted content. Critical story or gameplay information might have been presented at the problem level, and other levels will have to be reworked (and retested) to accommodate this.
- **which bugs to ship with.** In many ways, this is the toughest decision of all: which bugs to leave in the game.

Living with Bugs (For Now)

As a player, you might have encountered a bug in a retail game you have played. Your reaction might likely have been, "I wonder how the testers missed this one?" Chances are that they did not miss the defect. It is highly likely that during the course of the game's development a tester found the same bug you did and wrote it dutifully into the bug database. The ugly truth is not all bugs get fixed.

There will be times, especially late in the project, when the development team determine that they cannot (or will not) fix a particular bug. This can happen for a variety of reasons. Perhaps the technical risks involved in the fix outweigh the negative impact of the defect. Perhaps there is a workaround in place that the technical support team can supply to players who encounter the defect. Perhaps there simply is not enough time.

Whatever the reason, each project must have a quick and orderly process in place to determine which defects will be *waived*. That means the bug will not be fixed by the development team (at least in the current release cycle).

NOTE

The "waived" designation has many different names. Waived bugs can be known as "as is," ISV (In Shipped Version), DWNF (Developer Will Not Fix), or CBP (Closed by Producer). The worst of all possible names for waived status is "featured," which institutionalizes the cynical joke, "It's not a bug. It's a feature." Not surprisingly, more than one studio that repeatedly used "featured" to describe waived defects is now defunct, having released too many buggy games.

Cynicism, defeatism, and defensiveness have no place in the bug waiving process. Testers work very hard and want to feel as though their efforts matter to the project. However, developers work just as hard, and have a duty to ship the game on schedule. It is crucial that all parties involved maintain an understanding and respect for the role each plays in the overall project team.

Ideally, the senior members of the project team will meet regularly and often to discuss those bugs that the development team has requested to be waived. These can be flagged as "waive requested" or "request as is" in the Status or Developer Status fields of the bug database. The senior members of the project team (for example, the producer, executive producer, lead tester, and QA manager) can meet to evaluate each waive request and to discuss the costs and benefits of fixing it versus leaving it in the game (or at least until the next patch). Other team members, such as programmers or testers, should be available for these meetings as needed. This decision-making body is sometimes called the Change Control Board (CCB) or the Bug Committee.

FIGURE 5.5 Grand Theft Auto IV: One would reasonably expect Nico Bellec and his car not to be launched high in the air when driving into a swing set. Source: *Grand Theft Auto IV*. © Rockstar Games, Inc.

"The Swing Set of Death"

Video games are intended to be fun to play. Designers and developers spend months and years trying to create an immersive, meaningful, well-crafted experience for players. Sometimes, however, a bug can create accidental fun for the player, even though it may completely break the realism of the game.

In *Grand Theft Auto IV*, there is a particular swing set that can launch players and their cars high in the air. Known colloquially as the "Swing Set of Death" bug, this clearly should have been fixed by the development team before release, or at least fixed in a post-release patch. It remains in the game years after its 2008 release, and has been the subject of dozens of video compilations shared online. Why?

We can surmise one of two scenarios:

1. ***The bug was found, but it was too risky to fix.*** The bug occurs at the intersection of several of the game's core systems: collision for solid environmental objects, driving, progressive car damage, and general environmental physics. Making a change to any of these systems in order to fix the bug could possibly have created multiple new bugs elsewhere in the game.
2. ***The bug was found, and the designers thought it was fun.*** This bug is a perfect example of "breaking" the game. The expected result of running your car into a swing set should be damage to both the car and the swing set. Being launched many stories into the air is clearly an "unexpected result" that interrupts the immersive experience of playing a character in gritty urban crime drama. But is it fun? The designers likely thought so. The fans of this famous bug clearly think so.

Spectacular and amusing bugs like this sometimes require the development team to make hard choices: to fix or not to fix?

Patches, Updates, and Hotfixes

In some cases, where a post-release software update, or *patch*, is anticipated and budgeted for, a number of bugs will be designated for fixing after the game has been shipped (see "Post-release Testing" later in this chapter). Because current game consoles and mobile devices are Internet-capable and have some on-board means of storing data, game developers rely on patches as a means of extending their development schedule. Whereas with pre-Internet PCs and older game consoles such as the PlayStation® 2, a game had to be ready to ship *before* it was manufactured on game disks or cartridges, developers can now fix bugs up until the game ships or goes live, as long as they have the patch tested and live by the street date. Further updates, including bug fixes, gameplay tweaks, and even new features and content, can be released at any time over the life of a game. If a new bug is inadvertently introduced by a patch or update, you can make a further update (a *hotfix*) to address that new issue.

Once a bug has been waived, it is important to remind both the bug author and the test team that merely because the bug was waived does not mean that it was not a legitimate bug. Nor does it mean that they should not continue to find defects with the same level of diligence. *It is the duty of the test team to write up every bug they find, every time, no matter at what point in the production cycle they find it.* Defect reports supply the lead tester, project manager, and the business unit heads (marketing, sales, and product development) with the best information possible about the state of the game so that the best business decisions can be made.

A Note on *Cyberpunk 2077*

The decision to release a buggy game can baffle players, critics, and the industry at large. The simultaneous release in 2020 of *Cyberpunk 2077* across multiple hardware platforms caused a global sensation, but not in the way that publisher CD Projekt Red likely intended. The number and severity of graphical, functional, and platform-specific bugs was so high that it was shortly de-listed from the PlayStation store. (Needleman 2024) The developers remained hard at work on the game over the next several years, releasing dozens of patches and updates, and the game ultimately began to earn positive reviews. This extraordinary case is an outlier, however, and regularly releasing bug-infested games is not and never should be a viable model for either game development or publishing.

GOLD TESTING

Once the Beta test phase is over, the game should be ready for release or, in the case of a console or mobile title which must be tested and certified by a third party, final submission. The following entry criteria are typical for release testing:

1. All known Severity 1 bugs (crashes, hangs, major junction failures) are fixed.
2. Greater than 90% of all known Severity 2 bugs are fixed.
3. Greater than 85% of all known Severity 3 bugs are fixed.
4. Any known open issues have a workaround that has been communicated to Technical Support (or documented in an FAQ or readme.txt file).
5. Release-level performance has been achieved (i.e., a 60 fps frame rate).

Upon meeting your release criteria, the game is declared to be at "code lock." A brief, intense period of testing should now be performed on what everyone on the team hopes will be (but which may likely not be) the final build. Because the version of the game that is sent to be manufactured is traditionally known as the *gold master*, the final few versions tested are known as *gold master candidates* (GMCs) or *release candidates* (RCs).

At this point, the game looks and feels like a released, commercial game. It is up to the testers to serve as the last line of protection for both the players and the project team by sniffing out any remaining hidden defects that might have a significant impact on player satisfaction. This should be done by rerunning all of the test suits (or as many as time permits) one final time. In addition, a number of testers should be tasked with "breaking" the game one final time. Any remaining bug found during this final effort deemed too severe to be waived is called a *show stopper* because it causes the GMC to be rejected. A new GMC must be prepared with a fix for this new defect, and Gold testing must start all over again from the beginning.

Last-Minute Defects

Because the final stages of the project are so intense and pressure-laden, developers are likely to react very negatively to show stoppers. "Why are we [or you] just finding this now? Testing has been going on for months!" This cry is frequently heard from stressed-out managers presented with yet another show stopper. It is best for the test team to be patient with such outbursts and remember several inviolable truths of video game development:

1. There is seldom enough time in any project to find every bug.
2. Every time a programmer touches the code, bugs might be introduced.
3. Code changes accumulate over time, so that several iterative changes to different parts of the game might result in a bug showing up downstream from those changes.
4. Programmers are tired and prone to making more mistakes toward the end of a project.
5. Testers are tired and prone to missing more bugs toward the end of the projects.
6. Bugs happen.

In the case of a PC game, Web game, and games for other "open" platforms, the game's publisher or financing entity is the sole arbiter of whether to release the product. In these cases, once the Gold testing phase has been concluded, the game is ready for release. In the case of a console game, however, there is one final gatekeeper—the platform owner (for example, Nintendo, Microsoft, or Sony—who must certify the code. This final release testing process is known as *certification testing*.

Release Certification

The publisher will send a clean GMC to the platform manufacturer for final certification once the project team has finished gold testing. The platform manufacturer then conducts its own intensive testing on the GMC. This testing consists of two phases, which can happen concurrently or simultaneously. The *standards phase* tests the code against the Technical Requirements Checklist. The *functionality phase* tests the code for functionality and stability. The certification testers generally play the game all the way through at least once per submission. They sometimes find show stopper bugs of their own.

At the end of certification testing, the platform owner's QA team will issue a report of all the bugs they found in the GMC. Representatives of the publisher will discuss this bug list with the representatives of the platform owner and will mutually agree upon which bugs on the list must be fixed. (Sometimes, depending upon the platform, such a negotiation is impossible. The publisher *must* fix all the bugs on the list and resubmit a new release candidate.)

The development team is well advised to fix only those bugs on the "must fix" list, and to avoid fixing each and every minor bug in an effort to please the platform owner. Fixing more bugs than is absolutely necessary to win final certification only puts the code at risk for more defects, and the release schedule at risk for further delays.

Once the game has been resubmitted and certified by the platform manufacturer, it is "gold." The development and test teams can celebrate. The project is not over yet, however.

POST-RELEASE TESTING

Patches are a fact of life in modern video game development. Users do not like them but want them if they are available. Publishers do not like them, because they potentially add to the overall cost of the project. Developers do not like them, because they can be perceived as a tacit admission of failure. However, if the game was shipped with even one or two bad defects, either intentionally or inadvertently, it is time for a patch.

The upside of developing a testing a patch is that it allows the development team to revisit the entire list of waived bugs and last-minute design tweaks to further polish the game. Each additional bug fix or feature polish means more testing, however, and should be planned for accordingly.

Sometimes the development team will release more than one patch. In that case, the testing becomes more complicated, because interoperability must be tested. Each new patch must be tested to see whether it functions with both the base released game and with earlier patched versions.

LIVE TEAMS

So many games today are conceived less as closed-ended products than as "live" services that are constantly being updated. Beyond mere "patches," these post-release updates help to keep a game alive for months—and often years—after its initial release. Such post-release, live updates can be prompted for any of three main reasons:

1. The developer or publisher wishes to release new features, improvements to existing features, or refreshed content (such as a new "season" of multiplayer competition).
2. The developer or publisher is required by the operating system (OS) or platform owner to update the game to make it compatible with an updated version of the OS or platform (such as Android or Steam).
3. Further bug fixes.

The concept of a "live team" was made popular with the development of MMORPG games, where the success of the game itself depended upon thousands of contented players interacting in a shared virtual environment. MMORPGs "went gold" at retail outlets, but they also "went live" on the servers where players connected with each other in the virtual game world. Once the game was released, a substantial portion of the development team was retained to respond to the needs of players to keep the community thriving and the subscription revenue coming in. The "live team" would release patches to fix bugs, rebalance gameplay and push new content into the game world. That "live team" concept has been adopted by a growing number of developers as so many games depend upon a happy community of players for continued success. These games (many of them so-called "free to play" games) are continually updated (and tested) until they reach the end of their commercial life, which is generally determined by the players, not the developers. As long as these open-ended, "live team" games continue to make money, publishers are happy to support them.

For whatever the reason, it is important to understand that each update, like each chapter of DLC, should be treated as a new, distinct product in terms of test planning and execution. Although it is tempting in a "live" state to cut corners and push builds to the release server as quickly as possible, careful testers will not allow the complexity of the development environment to become an excuse for not doing their jobs carefully and according to a written test plan based on a written specification of the update, no matter how trivial it may seem.

A Note on DLC

DLC, or *downloadable content,* is growing more popular (and is even expected) with players. Sometimes DLC is planned after a game's release, and sometimes it is planned during its development. DLC can take many forms and sizes, from additional vehicles or outfits, to map packs or bonus levels, to completely new storylines and casts of characters. In items and levels can be purchased à la carte, and are released regularly throughout the game's life cycle.

No matter what its size, *each DLC release should be treated as a new product* and subject to all the planning, test kickoffs, and phase entry criteria described in this chapter. DLC should not be marginalized or treated as less of a product, nor should it be tested with less diligence, merely because it is released after its parent game.

Structured game testing breaks the test activities into distinct phases, each of which has its own inputs, deliverables, and success criteria. These phases reflect the progressive completion and improvement of the game code until it is finally fit to be released to the playing public. Once test planning and preparation are completed, different types of testing are used in the remaining phases. Like pieces of a mosaic, they each reveal something different about the game code—in the right place and at the right time.

EXERCISES

1. What are the main responsibilities of a lead tester?
2. The Beta build is the version that will be sent to be released to the playing public. True or False?
3. Describe whether each of the following is an appropriate topic to discuss during a test execution kickoff, and why:
   a. Possible contradictions in the feature requirements
   b. Ideas for new tests
   c. Company stock prices
   d. Identical tests already being run in other test suits
   e. How "buggy" a feature was in a previous release
   f. Recent changes to the game data file formats
   g. Lack of detail in the test case documentation
4. All bugs must be fixed before a build can be considered a GMC. True or False?
5. Explain the difference between a test case and a test plan.
6. The QA lead on your live team leaves the company, and you are promoted to take their place. Your first assignment is to test the next content update of your game. What phases will you use in planning your update testing?

REFERENCE

Needleman, Sarah. 2024. "How a $300 Million Flop Turned Into an Improbable Hit." *The Wall Street Journal*, January 11, 2024. *https://www.wsj.com/tech/cyberpunk-2077-videogame-flop-to-hit-58ed2741*.

CHAPTER 6

EXPLORATORY TESTING AND GAMEPLAY TESTING

Although most of this book is designed to help you take a methodical, structured approach to testing a game, this chapter focuses on more intuitive, unstructured—yet no less crucial—approaches to game testing. *Exploratory testing*, sometimes called *ad hoc testing* or *general testing*, describes searching for defects by exploring the game as a player might do. *Gameplay testing* describes playing the game to test for such subjective qualities as balance, difficulty, and "fun factor."

AD HOC TESTING

Ad hoc is a Latin phrase that can be translated as "to this particular purpose." Exploratory testing is, in its purest form, a single test improvised to answer a specific question.

Despite the most thorough and careful test planning and test design, or the most complex test suite you might have developed, even after being reviewed carefully by other test leads or the project manager, there is always something you (or they) might have missed.

Exploratory testing allows you, as an individual tester, to pursue investigative paths that perhaps occurred to you, even unconsciously, in the course of performing structured test suites on the game. During the course of testing a game you will, almost daily, have thoughts along the lines of, "I wonder what happens if I do…?" Exploratory testing gives you the opportunity to answer such questions. It is the mode of testing that best enables you to explore the game, wandering through it as you would a maze.

There are two main types of exploratory testing. The first is *free testing*, which allows the professional game tester to "depart from the script" and to improvise tests on the fly. The second is *directed testing*, which is intended to solve a specific problem or to find a specific solution.

Free Testing Comes from the Right Side of Your Brain

Because it is a more intuitive and less structured form of testing, free testing is sometimes called "right-brain testing." Nobel Prize-winning psychobiologist Roger W. Sperry asserted that the two halves of the human brain tend to process information in very different ways (Sperry 1968, 723). The left half of the brain is much more logical, mathematical, and structured. The right half is more intuitive, creative, and attuned to emotions and feelings. It is also the side that deals best with complexity and chaos.

As the video game industry continues to grow, there is continued pressure for "bigger, better, and more" in every aspect of a game's design—more features, more user customization, more content, more genre-blending, and more complexity. At its best, exploratory testing allows you as a tester to explore what can, at times, appear to be an overwhelming game design.

Exploratory testing also presents you with an opportunity to test the game as you would play it. What type of game player are you? Do you like to complete every challenge in every level and unlock every unlockable? Do you like to rush or build up? Do you favor a running game or a passing game? Do you go quickly through levels or explore them leisurely? Free testing allows you to approach the game as a whole and to test it according to whatever style of play you prefer. (For an expanded discussion of player types, see Chapter 10, "Cleanroom Testing.")

"Fresh Eyes"

Fatigue, carelessness, and apathy are all enemies of good game testing. Those testers who must exercise the same part of a game repeatedly are most at risk, but over the course of a long project, sooner or later, each team member is likely to suffer from one condition or another. It is easy for testers to become *snow blind*, a condition in which you have been looking at the same assets for so long that you can no longer recognize anomalies as they appear. You need a break. You need to do something different.

! TIP

Test With Your Ears!
Even though you may have been testing the same game for days on end, avoid listening to music or podcasts while testing a game. You never know what new audio bugs may have been introduced (or that you might have just discovered) in the current build.

Exploratory testing can allow you to explore modes and features of the game that are beyond your primary area of responsibility. Depending on the manner in which your project is managed, you might be assigned to one specific area, mode, feature, or section of the game. The test suites you perform on each build might focus only on that specific area. Ad hoc testing allows you to move beyond to other areas, and allows other testers to explore your area, without a test suite to guide them.

This method can include the following:

- assigning members of the multiplayer team to play through the single-player campaign
- assigning campaign testers to skirmish or multiplayer mode
- assigning the config/compatibility/install tester to the multiplayer team
- assigning testers from another project entirely to spend a day (or part of a day) on your game
- asking non-testers from other parts of the company to play the game (see the section "Gameplay Testing" later in this chapter)

Who Turned the Lights On?

A venerable PC games publisher operated a handful of test labs in its various studios around the country, and the local test managers would often send builds of their current projects to each other for ad hoc testingand final checking.

When one test manager handed the latest build of another studio's Formula One-type racing game to two of his testers, he was surprised to see them back in his office minutes later.

"Crashed it already!" they proudly reported.

"How?" the manager cried. "You've barely had time to finish a race!"

"We just turned the headlights on!"

As you might expect, the default time in the default track in the default mode was "day." When the two testers started their race in this mode, they turned their cars' headlights on "just to see what happens." The game crashed instantly.

Needless to say, this counterintuitive pair of settings (`time = day` and `headlights = on`) was added to the combinatorial tables by the chastened, but wiser, test lead.

Directed Testing Makes Order Out of Chaos

Ad hoc testing is a natural complement to structured testing, but it is by no means a substitute for it. Whether you have been given a specific assignment by your test lead or you are playing through the single-player campaign "just to see what happens," your testing should be documented, verifiable, and worthwhile.

Set Goals and Stick to Them

Before you begin, you should have a goal. It need not (and should not) be as complex or as well thought out as the test cases and test suites discussed later in the book. You need to know where you are going so you do not waste your (and the project's) time, however. Briefly write out your test goal before you launch the game.

This goal can be very simple, but it must be explicit. Here are some examples:

- Can I play a full game by making only three-point shots?
- Is there a limit to the number of turrets I can build in my base?
- Can I deviate from the strategy suggested in the mission briefing and still win the battle?
- Is there anywhere in the level I can get my character stuck?
- Can I buy a unique item more than once?

NOTE *Whether you actually achieve the goal of your free testing is less important. If, in the course of trying to reach your goal, you stumble upon a defect you had not intended to find, that is fine. That is what free testing is all about.*

Manage Your Multiplayer

If you are leading a multiplayer test, let all the other testers know the purpose of the game session before it starts. Successful multiplayer testing requires communication, coordination, and cooperation, even if it seems that the testers are merely running around the level trying to kill each other. In most circumstances, one tester should direct all of the other players in order to reach an outcome successfully, which can often be difficult. If even one tester in a multiplayer test loses sight of the aim of the test, the amount of time wasted is multiplied by the number of testers in the game. Do not let your team fall into this trap.

Don't Say "Play!"

In your testing career, avoid the use of the verb "to play" when you refer to game testing. This will help to counter the widely held notion that the QA team "just plays games for a living." It will also help to reinforce to your test team that your work is just that, work. Remember: the first time you play through the game, you are playing. The fortieth time, you are working.

If You Are Not Recording, You Are Not Testing

You should continually take notes as you are testing through a game. Game designer Sid Meier (*Civilization*) said that good games are "a series of interesting choices" (Rollings & Morris 2000). It is imperative that you keep track of these choices by writing down which options you choose, paths you take, weapons you equip, and plays you call in a very meticulous and diligent manner. In so doing, when you encounter a defect, you will be better able to come up with a reproducible path.

Documentation could be difficult when you are in the middle of a 12-trick chain in a *Tony Hawk*-style stunt game. That is where video capture becomes an almost indispensable test tool. In designing your test configuration, allow for some "minimally invasive" video recording solution that will not affect the performance of the game on the target hardware. Many modern game consoles have built-in video and screenshot capture tools. If you are testing on tablet or smartphone, it may be necessary to rig a video camera on a tripod to record gameplay on the target device.

FIGURE 6.1 *Marvel Snap* (iOS): Do not be afraid to annotate the screenshots or videos you include in a bug report. Source: *Marvel Snap*. ©MARVEL ©Nuverse (Hong Kong) Limited ©Second Dinner

PC games can be captured with such third-party software tools as Bandicam (*www.bandicam.com*), Fraps (*www.fraps.com*), Camtasia Studio (*www.techsmith.com/camtasia*), or Game Bar (*https://apps.microsoft.com/detail/9NZKPSTSNW4P*). The drawback to using video capture software, however, is that you risk tasking the computer's resources during the game's runtime, thereby possibly creating defects or lowering performance benchmarks than you would normally experience without the video capture software running simultaneously on the system. As part of the test planning phase, lead testers should work with both hardware and software engineers to arrive at a "code friendly" solution that all parties are confident will not introduce false defects.

Testing video should not become a crutch, or an excuse for less-than-diligent work on the part of the tester. It should serve as a research tool and a last-resort means of reporting a defect. Use the following steps as a guide:

1. Start the DVR (or capture software) and press the record button before you start the game. (It is too easy to forget, otherwise.)
2. When you come to a defect you cannot reproduce, rewind the recording, study it, and then show it to your test lead and colleagues to discuss what could have caused the bug and whether anyone else has seen the same behavior in similar circumstances.
3. If you absolutely, positively, cannot reproduce the defect, copy a clip of the video and attach it to the bug report, email it to the developer, or copy it to a project folder for future reference.
4. Once you have filled up the DVR, archive the captured video. Video files tend to be very large, so you should establish some network backup protocol to prevent your local hard drive from filling up with gameplay captures.

Free testing should have clear goals. The work done should be documented (via video) and documentable (through clear, concise, reproducible bug reports). It should also be worthwhile. The following are but a few of the common pitfalls you should avoid when free testing:

- Competing with other testers in multiplayer games. The job is not about getting the highest score or the most wins. It is about finding bugs and delivering a good game.
- Spending a great deal of time testing features that could be cut. You might be made aware that a certain mode or feature is *on the bubble*, that is, in danger of being eliminated from the game. Adjust your free testing focus accordingly.
- Testing the most popular features of the game. Communicate frequently with your test lead and colleagues so you can stay current with what areas, features, and modes have been covered (and re-covered) already. Focus your time on the "unexplored territory" of the game.
- Spending a disproportionate amount of time testing features that are infrequently used. You could be wasting time spending day after day exploring every nook and cranny of the map editor in your RTS, for example. Only about 15% of all users typically ever enter a map editor, and fewer than 5% actually use it to create maps. You want those players to have a good experience, but not if it places the other 85% of your players at risk.

Avoid Groupthink

Because ad hoc testing depends on the instincts, tastes, and prejudices of the individual tester, it is important as a test manager to create an environment in which testers feel free to think differently from one another. Game players are not a uniform, homogeneous group; your test lab should not be, either. If you have staffed your lab with nothing but self-identified "hardcore" players, you will not find all the bugs, nor will you ship the best game.

Groupthink is a term coined by social psychologist Irving Janis to describe a situation in which flawed decisions or actions are taken because a group under pressure often sees decay in its "mental efficiency, reality testing, and moral judgment" (Janis 1972). One common aspect of groupthink is a tendency toward self-censorship: individuals within a group fail to voice doubts or dissent out of a fear of being criticized, ostracized, or worse. This can be a danger in game testing because people who seek tester jobs tend to be in their very late teens or early 20s, young enough that pressure to conform to their peer group is still very strong.

TIP

Be a "Hardcore" Tester!
Hardcore gaming is not the same as hardcore testing. Hardcore gamers are generally used to tolerating mistreatment: they willingly pay for games weeks and months before they are even released; they gladly suffer through launch week server overload problems; and they love to download patches. Use the methods described in this book to get them to understand that bug-fixing patches can be the exception, rather than the rule. All it takes is careful test planning, design, and execution.

You may encounter attitudes in your test lab such as:

- "Everybody has broadband, so we don't need to test modem play."
- "Nobody likes the L.A. Clippers, so I won't test using them as my team."
- "Everybody played *StarCraft*, so we don't need to test the tutorial in our own RTS."
- "Nobody likes CTF (capture the flag) mode, so we don't need to spend much time on it."
- "Nobody uses melee weapons, so I'll just use guns."

Your job as a tester, and as a test manager, is to be aware of your and your team's particular biases, and to create an atmosphere in which a variety of approaches—to both playing and testing—are discussed and respected freely and frequently. Cultivate and encourage different types of play styles. Recruit sports gamers. Recruit casual and non-gamers. Recruit older testers. Recruit professionals who understand that this is a skilled job, not a hobby.

Testing as Detective Work

The second broad category of ad hoc testing is *directed testing*. You could best describe this method as "detective testing," because of its specific, investigative nature. The simplest form of directed testing answers very specific questions, such as:

- Does the new build work?
- Can you access all the characters?

- Are the cut scenes interruptible?
- Is saving still broken?

The more complex type of directed testing becomes necessary when testers find a major defect that is difficult to seemingly impossible to reproduce. The tester has "broken the game," but cannot figure out how they did it. As with a good homicide case, the tester has a body (the bug) and an eyewitness (the tester). Unlike a homicide case, the focus is not on who created the bug: the perpetrator is the defect somewhere in the code. The focus is on "How did the defect happen?" Answering this question will help the programmers to find the exact location of the culprit.

Directed testing commonly begins when one or more testers report a "random" crash in the game. This can be very frustrating, because it often delays running complete test suites and a significant amount of time could be spent restarting the application and re-running tests. Unstable code, especially in the later phases of the project, can be stressful. You will need to be patient and think clearly so that you can do your best detective work.

! TIP

"Everything Happens for a Reason"
So-called "random crashes" are seldom truly random. Use directed testing and the scientific method to eliminate uncertainty along your path to reproducing the bug often enough so that you can get the development team to find it and fix it.

The Benefits of Reproduction

One of the most critical bits of information in any bug report is the rate of reproduction. In a defect tracking database, this field might be called (among other things) frequency, occurrence rate, "happens," or "repro rate." All these various terms are used to describe the same thing. *Reproduction rate* can be defined as the rate at which, following the steps described in a bug report, anyone will be able to reproduce that bug.

The word "anyone" is critical to the above definition of reproduction rate. A defect report is not very helpful if the tester who found the bug is the only one able to re-create it. Because video game testing is often skill-based, it is not uncommon to encounter a defect in a game (especially a sports, fighting, platforming, or stunt game) that can only be reproduced by one tester, but that tester can reproduce the bug 100% of the time. In an ideal situation, that tester will collaborate closely with other members of the team so that they can identify a path that will allow the others to re-create the bug.

If this is not possible due to time or other resource constraints, be prepared to send a video clip of the defect to the development team or, in the worst case, send the tester to the developer to do a live demonstration of the bug. This can be costly and time-consuming because, in addition to any travel expenses, the project is also paying for the cost of having the tester away from the lab (and not testing) for a period of time.

The reproduction rate of a defect is generally expressed as a percentage ranging from 100% to "once," but this can be misleading. Assume, for example, that you find a defect during the course of your free testing. After a little research, you narrow down the steps to a reproducible path. You follow those steps and get the bug to happen again. You could, reasonably, report the defect as occurring 100% of the time—you tried the steps twice and it happened both times. It could be just as likely that the bug is only reproducible 50% of the time or less, however, and you just got lucky, as though you had flipped a penny and got it to land heads-up twice in a row.

For this reason, many QA labs report the reproduction rate as the number of observed occurrences over the number of attempts (for example, "8 out of 10"). This information is far more useful and accurate, because it allows your test lead, the project manager, and anyone else on the team to evaluate how thoroughly the bug has been tested. It also helps to keep you honest about the amount of testing you have given the defect before you write your report. How fair is it for you report that a crash bug happens "once" if you tried to reproduce it only once? If you want to maintain your credibility as a member of the test team, you will not make a habit of this.

With certain defects, however, even a relatively novice tester can be certain that a bug occurs 100% of the time without iterative testing. Bugs relating to fixed assets, such as a typo in in-game text, can safely be assumed to occur 100% of the time.

In summary, the more reproducible a bug is, the more likely it is that it will be fixed.

The Scientific Method

It is no coincidence that the room where game testers work is often called a lab. Like most laboratories, it is a place where the scientific method is used both to investigate and to explore. The scientific method consists of the following steps:

1. Observe some phenomenon.
2. Develop a hypothesis as to what caused the phenomenon.
3. Use the hypothesis to make a prediction; for example, "If I do this, it will happen again."
4. Test that prediction by retracing the steps in your hypothesis.
5. Repeat Steps 3 and 4 until you are reasonably certain your hypothesis is true.

These steps provide the structure for any investigative directed testing. Assume you have encountered a quirky defect in a PC game that seems difficult to reproduce. Perhaps it is a condition that either breaks a script, gets your character stuck in the geometry of the level, causes the audio to drop out suddenly, or causes your PC to crash. Here is what to do:

First, review your notes. Quickly write down any information about what you were doing when the defect occurred while it is still fresh in your mind. Review the video. Determine as best as you can the very last thing you were doing in the game before it crashed.

Second, process all this information and make your best educated guess as to what specific combination and order of inputs could have caused the crash. Before you can retrace your steps, you have to determine what they were. Write down the input path you think most likely caused the crash.

Third, read over the steps in your path until you are satisfied with them. You guess that if you repeat them, the defect will occur again.

Fourth, reboot your computer, restart the game, and retrace your steps. Did you get the crash to occur again? If you did, great!

Fifth, write it up. If you did not re-create the bug, change one (and only one) step in your path. Try the path again, and so on, until you successfully re-create the defect.

Unfortunately, games can be so complex that this process can take a very long time if you do not get help. Do not hesitate to discuss the problem with your test lead or fellow testers. The more information you can share, the more brainstorming you can do, the more "suspects" you can eliminate, and the sooner you will find the bug.

GAMEPLAY TESTING

Gameplay testing (or *play testing*) is entirely different from the types of testing discussed so far in this book. Other chapters in this book are concerned with the primary question of game testing: Does the game work as designed? Play testing concerns itself with a different but arguably more important question: Does the game work *well*?

The difference between these two questions is obvious. The answer to the first question is binary; it is either yes or no. The answer to the second question is far from binary because of its subjective nature. It can lead to many other questions:

- Is the game too easy?
- Is the game too hard?
- Is the game easy to learn?
- Are the controls intuitive?
- Is the interface clear and easy to navigate?

The most important question of all is

- Is the game fun? (For what types of players?)

Unlike the other types of testing covered thus far, gameplay testing concerns itself with matters of judgment, not fact. As such, it is some of the most difficult testing you can do.

A Balancing Act

Balance is one of the most elusive concepts in game design, yet it is also one of the most important. *Balance* refers to the game achieving a point of equilibrium between various—usually conflicting—goals:

- challenging, but not frustrating
- easy to get into, but deep enough to compel you to stay

- simple to learn, but not simplified
- complex, but not baffling
- long, but not too long

Balance can also refer to a state of rough equality between different competing units or factions in a game:

- melee fighters vs. ranged fighters
- rogues vs. warlocks
- sniper rifles vs. rocket launchers
- the Covenant vs. Humanity
- Ken vs. Ryu
- the Plants vs. the Zombies

The test team might be asked by the development team or project manager for balance testing at any point in the project life cycle.

! TIP

Balance at Beta
It is wise to save any serious balance testing until at least Beta because it is difficult to form useful opinions about the interplay of different game systems until they are all implemented and working together in the code. This typically happens at the Beta stage.

Once the game is ready for gameplay testing, it is important for test feedback to be presented in as specific, organized, and detailed a manner as any other defect report. Some project managers may ask you to report such balance issues as bugs in the bug database; others may ask the test lead to collect gameplay and balance feedback separate from defects. In either case, express your gameplay observations so that they are presented as based in fact, and hence, sound credible.

Let's examine some feedback collected from testers during balance testing on *Battle Realms*, a PC real-time strategy game (RTS) developed by Liquid Entertainment. It became clear very early in the course of play testing that the Lotus Warlock unit was overpowered (or *OP*). One tester wrote:

```
Lotus Warlocks do too much damage and need to be nerfed.
```

If you have spent any time on Internet message boards, comments like this should look very familiar. The tester is not specific. How much damage is too much? Relative to what? If *nerfed* means "made less powerful," how much less? 50%? 50 points? The development team is not very likely to take this comment seriously, thinking it is an impulsive, emotional reaction. (It so happens that it was. The tester had just been on the receiving end of a warlock rush.)

```
Lotus Warlocks should have a five-second cooldown added
to their attack.
```

This tester is overly specific. He has identified a problem (overpowered warlocks) and gone too far by appointing himself game designer and declaring that the solution is a five-second cooldown (that is, a delay of five seconds between the end of a unit's attack and the beginning of its next attack). This comment presumes three things: that the warlocks are indeed overly powerful, that the designers agree that the best solution is to implement a cooldown, and that the code has been written (or can be written) to support a cooldown between attacks. The development team is likely to bristle at this presumption (even if it is a viable solution).

```
Lotus Warlocks are more powerful than the other three
races' highest-level unit. Their attack does approximately
10% more damage than the Dragon Samurai, Serpent Ronin,
and Wolf clan Werewolf. They get three attacks in the
same time it takes the other clans' heavy units to do two
attacks. Players who choose to play as the Lotus Clan win
75% of their games, frustrating the non-Lotus players.
```

This comment is specific and fact-based. It gives the producers and designers enough information to start thinking about rebalancing the units. It does not, however, suggest how the problem should be solved. Devising a solution to the problem is the job of the game designers.

Interview: Karen McMullan

Determining what emotions a player is feeling during a game is a very important part of play testing, according to Karen McMullan, who worked as a content designer on some of Ensemble Studio's biggest games (among them *Age of Mythology*, *Age of Empires III*, and *Halo Wars*).

"The most useful thing for me as a designer is for you to tell me what you're feeling. What you're thinking about. What decisions you made, and why." Ms. McMullan suggests expressing gameplay feedback by "leading with a feeling and following it up with a reason. 'I was frustrated because my spearmen lost to chariots. Infantry are supposed to beat cavalry, right?' for instance."

(Permission: Karen McMullan)

"It's Just a Suggestion"

Play testing occurs constantly during bug testing. Because testers are not robots, they will always be forming opinions and making judgments, however unconscious, about the game they are testing. Occasionally, a tester will feel inspired to suggest a design change. In some labs, these are called *suggestion bugs* (or *severity* S bugs), and they are frequently ignored. Because bugs create stress for programmers, artists, and project managers, they rarely appreciate the database being cluttered up with suggestion bugs.

A far more successful method of making your voice heard as a tester, if you are convinced that you have a valuable (and reasonable) idea for a design change, is the following:

1. Ask yourself whether this is a worthwhile change. "Zorro's hat should be blue," is not a worthwhile change.
2. Get a good night's sleep before mentioning the suggestion. It might not seem like such a good idea in the morning.
3. Express your idea in the positive. "The pointer color is bad," is a far less helpful comment than, "Changing the pointer to green would make it easier to see."
4. Discuss it with your fellow testers. If they think it is a good idea, then discuss it with your test lead.
5. Ask your test lead to discuss it with the project manager or lead designer.
6. If your test lead convinces the development team that your idea has merit, at that point you might be asked to enter the suggestion into the defect database as a bug so that it can be tracked like any other change. Only do this if you are asked to do so.

Testers can often have their suggested design tweaks incorporated into games by using this process: discussing the idea, getting the team to agree with the suggestion, and communicating it to the developers outside of the bug database. Know your role, and work within the system.

Making a Game Easy is Hard Work

One element of game balance that becomes the most challenging late in the development cycle is the game's difficulty. A new game should be "easy" enough so that new players feel welcomed, but "hard" enough so that new players will feel rewarded by taking on greater challenges. The problem with balance testing is that regular game testers become experts at the game, so they are poor judges of the game's difficulty.

Games take months and years to develop. By the time a game enters full-bore testing, the game testers will likely have completed the game more often than even the most ardent fan. The design and development team might have been playing the game for more than a year. Over the course of game development, the following take place:

- Skills improve with practice. If you could not grind a rail for more than 10 feet when you got the first test build of an action sports game, you can now grind for hours and pull off 20-trick combos without breaking a sweat.
- AI patterns, routes, and strategies are memorized. The behaviors of even the most sophisticated NPC opponents become predictable as you spend weeks playing against them.
- Puzzles stop being puzzling. In adventure games or other types of games with mazes or hide-and-seek elements, once you learn how to solve a puzzle or where an item is hidden, it is impossible to unlearn it.
- Tutorials stop tutoring. It is very difficult to continue to evaluate how effective a lesson is if you have already learned the lesson.
- Jokes become stale.
- What was once novel becomes very familiar, almost boring. (See the discussion of "Fresh Eyes," earlier in this chapter.)

The upside of all this is that, on release day, the development and test teams are the best players of their own game on the planet. (This will not last long, though.)

The downside is that you (and the rest of the development team) lose your ability to objectively evaluate difficulty as the game approaches release. Nothing of what is supposed to be fresh and new to a player seems fresh and new to you. That is why you may need another set of "fresh eyes:" outside gameplay testers.

External Testing

External testing begins with resources outside of the test and development teams, but still inside your company. These opinions and data can come from the marketing department, as well as other business units. It is a good idea to have everyone who is willing, from the CFO to the part-time receptionist, gameplay test the game if there are questions that remain to be answered.

Here you must be careful to keep in mind the "observer effect," which describes the phenomenon that people may behave differently if they know they are being watched. Even small children may be aware that they are participating in a focus group or a play test. Because they (or their adult counterparts) are often eager to please the researchers, they might tell you what they think you want to hear, rather than what they really feel. Entire books have been written on how to manage this problem with consumer research (Sudman 2002).

Although outside gameplay testing and opinion gathering is an effort typically initiated by the development or design teams, it is often implemented and managed by the test team.

Subject Matter Testing

If your game takes place in the real world, past or present, the development team will perhaps wisely choose to have subject matter experts (or *content testers*) review the game for accuracy. See the sidebar, "Testing for Realism," about how real-life input from experienced fighter pilots enhanced the development of a game.

Testing for Realism

During the development of the PC jet fighter simulator *Flanker*, producers at the publisher, SSI, used the Internet to recruit a small group of American and Russian fighter pilots who were given Beta builds of the game. Their feedback about the realism of the game, from the feel of the planes to the Russian-language labels on the cockpit dials, proved invaluable.

These experts posted their comments to a password-protected message board, and their feedback was carefully recorded, verified, and passed on to the development team. The game was released to very good reviews and was given high marks for its realistic depiction of Soviet-era fighter planes.

Such an expert panel tends to be relatively small and easy to manage. It is much more challenging to manage a mass Beta test effectively.

External Beta Testing

External Beta testing can give you some very useful data. It can also give you an unwieldy amount of useless data if the testing is not managed properly. Beta testers are a very special component of the QA team. They provide a "volunteer army" of testers made up of game players who sign up to donate their time to playing pre-release Beta versions of a game. The size of the Beta testing team can range from several dozen to several hundred or more. They can expect to endure severe game problems and frequent downloads of new game builds or patches. For this privilege, Beta testers commit to documenting any problems they find so that the game programmers can have the opportunity to fix those defects prior to the official release of the game.

There are two types of Beta testing: closed and open. *Closed Beta* occurs first and is tightly controlled. Closed Beta testers are screened carefully and usually have to answer many questions before they are accepted into the Beta test. These questions can range from the technical specifications of their computer to which specific games they have played recently. After the testers play the game, they are asked to complete an online questionnaire or to participate in a message board discussion. They could also be invited to report any bugs they might find.

Open Beta occurs after Closed Beta concludes. Open Beta is open to all who are interested in participating. Although developers will still solicit some level of gameplay feedback from this much larger group, their role is to load test the network code and to identify issues with items such as the login system, matchmaking, overall network stability, and the in-game economy.

Although Open Beta testers will not run test cases, they could report defects, in addition to providing gameplay feedback. Most Beta test managers will host a bug reporting site that allows Beta testers to report defects, make comments, and ask questions.

Becoming a Beta Tester

Besides playing the game the way it would "normally" be played, here are some other strategies you can adopt as an individual Beta tester:

- Try to create infinite point-scoring, money-making, or experience-producing strategies.
- Try to find ways to get stuck in the game environment, such as a pinball bouncing forever between two bumpers or an adventurer who falls into a river and cannot get out.
- Spend some time in every feature, mode, or location provided in the game.
- Spend all of your time in one feature, mode, or location, and fully explore its options and functions.
- Try to find ways to access prohibited modes or locations.
- See what happens when you try to purchase, acquire, or use items and abilities that were designed for characters at a level much higher than yours.
- Try to accomplish something "first" in the game, such as becoming the first "maxed-out" character, the first to enter a particular town, the first to win a match, or the first to form a clan.
- Wear, wield, and activate as many stat-increasing items as you can at one time, such as armor or power-ups.
- Try to be the player with the "most" of something in the game, such as wins, points, money, trophies, or vassals.

For their efforts, Beta testers can be rewarded in various ways, such as:

- getting special items to use in the released game
- getting first choice for particular character names
- swag, either in-game, or in real-life, like a t-shirt or hat
- getting their name listed in the game credits

Becoming a Beta tester is a good way for someone who does not have formal training or game industry experience to build a résumé of accomplishments. You can visit *https://massivelyop.com/tag/betawatch/* as well as game company websites to check for the most recent Beta test and opportunities.

WHO DECIDES?

Ultimately, decisions that relate to changing the design, rebalancing, adding (or cutting) features, or even delaying the release to allow more time for "polish," are not made by game testers. Their role is to supply the appropriate decision makers and stakeholders with the best facts and advice they can about current the condition of the game, so they can make the best business decisions.

As your experience as a game tester grows, you may find opportunities to move into management as a primary tester, test lead, or even a QA manager or project manager. The next several chapters will help you to think about test creation and management as you move your career forward.

EXERCISES

1. It is a good idea to keep the same tester performing the same test cases for the length of a project. True or False?
2. Why is it unwise for game testers to refer to the work they do as "playing?"
3. Discuss the differences (in both method and results) between free testing and gameplay testing.
4. What are two methods of expressing a defect's reproduction rate?
5. You and seven other testers jump into a deathmatch session of the online shooter game you are testing. Once the game starts, it is a free-for-all, with each tester competing to win the session. Is this gameplay testing or ad hoc testing? Why?
6. You have been assigned to test the gameplay of an *Injustice*-type fighting game and suspect that one of the fighters seems significantly weaker than the others. What exploratory tests can you perform to confirm and quantify your suspicion?

REFERENCES

Janis, Irving Lester. 1972. *Victims of Groupthink: A Psychological Study of Foreign-policy Decisions and Fiascoes*. Boston: Houghton, Mifflin.

Rollings, Andrew and Dave Morris. 2000. *Game Architecture and Design*. Scottsdale, Arizona: Coriolis.

Sperry, R. W. 1968. "Hemisphere deconnection and unity in conscious awareness." *American Psychologist*, 23(10), 723-733. *https://doi.org/10.1037/h0026839*.

Sudman, Seymour and Brian Wansink. 2002. *Consumer Panels*. United States: American Marketing Association.

CHAPTER 7

The Two Rules of Test Management

RULE ONE: DO NOT PANIC

In a game project, panic is bad. The person panicking did not choose to panic, and may not realize it is happening. It is an irrational reaction to a set of circumstances, and it can lead a tester to cause harm to the project. When a tester is reacting inappropriately to some unreasonable request, remind him not to panic by asking "What's rule number one?"

Scuba divers put themselves in situations similar to what game testers might face: limited resources (the equipment you bring with you), time constraints (air supply), rules to follow (rate of ascent/descent), and other surprises (unexpected sea visitors). Episodes of panic may explain recreational diving accidents and deaths. Panic attacks can be triggered by something serious: entanglement, an equipment malfunction, or the sight of a shark. However, panic can make situations *worse* by leading to irrational and dangerous behavior.

Testing the wrong build, failing to notice an important defect, or sending developers on an avoidable search for a nonexistent bug should not get you physically hurt, but there will be a price to pay in extra time, extra money spent, or loss of sales and reputation.

Game project panic happens when you are:

- unfamiliar
- unprepared

- under pressure
- unrested
- "nearsighted"

Unfamiliar

As a member of the game team, you might be asked to do something you have never had to do before. You might be given someone else's tests to run, be placed into the middle of a different game project, or be told to take someone else's place at the last minute to do a customer demonstration. In situations like these, rely on what you know, stick to basics, and learn new or different approaches by watching the people who have already been doing them.

You might even be asked to accomplish something you have never done before, such as achieving 100% automation of the installation, or writing a tool to verify the foreign language test in the game. May be *no one* has ever done this before. Do not make a commitment right away, do not invent scenarios, and do not try to be a hero. If you are unfamiliar with a situation, you might act on your best judgment, but it still might not be right. A novel, stressful situation requires that you know when to get help, as well as some humility so you do not feel like you have to take on everything yourself or say "yes" to every request. You do not need to lose any authority or credibility. Find someone who has been in this situation before and who might help you discover some working solutions. Stay away from responses that are known to fail. You can even search the Internet to see if anyone else has been put in the same situation and is willing to share their experience.

Unprepared

A number of unexpected events will happen on your project. Expect the unexpected! Many parts of the game need to be tested at various points in the game's life cycle. Behind the scenes, many different technologies are at work. If you are not ready for a variety of test assignments and you do not have the skills needed to perform them successfully, then you will fail.

Study, practice, and experience are ingredients for good preparation. During the course of the project, try to get to know more about the game code. Keep up with the video game industry so you are also aware of what the next generation of games and technologies will be like. Become an expert in requirements and designs for the parts of the game you are responsible for testing, and then get familiar with the ones that you *are not* responsible for. When you least expect it, you might need to take on a different position, fill in for another tester, or grow into more responsibility. Be ready when it happens.

Under Pressure

Pressure can come from any of three directions:

- schedule (calendar time to complete the project)
- budget (money to spend on the project)
- headcount (the quantity and types of people assigned to work on the project)

There's nothing to prevent one or more of these resources from shrinking at any time during the project. These factors will not be under your control as a tester. Usually, they are determined by business conditions or project managers. In any case, you will be affected. Figure 7.1 shows the resources in balance with the scope of the project.

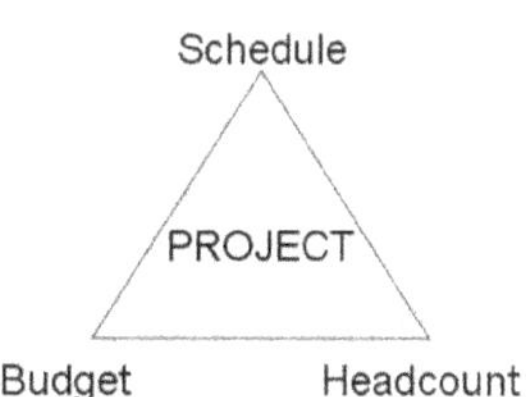

FIGURE 7.1 Resources balanced with the project scope

Moving in any one of these points on the triangle squeezes the project, creating pressure. Sometimes a game project starts out with one of these factors being too small, or they can get smaller any time after the project has launched. For example, money can be diverted to another game, developers might leave to start their own company, or the schedule gets pushed up to release ahead of a newly announced game that competes with yours. Figure 7.2 shows how a budget reduction can put pressure on the game project's schedule and headcount.

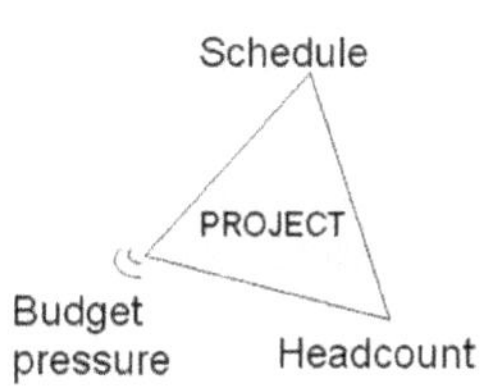

FIGURE 7.2 Budget reduction causes pressure

Another way to cause pressure within this triangle is to add more to it than it was originally planned for. This demand could be internally driven, such as adding more levels or characters, or scrapping the old graphics engine for a new one to take advantage of newly announced hardware.

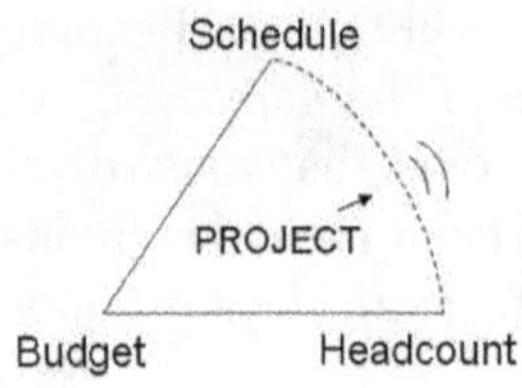

FIGURE 7.3 Budget and headcount pressure caused by scope increase

Other unplanned changes might be made up to support more game platforms than originally planned or to keep up with competing games in terms of levels, characters, online players supported, and so on. Figure 7.3 illustrates how increasing the scope of a project can put pressure on the budget and headcount if they are not also increased.

When there is pressure on the project, you can expect it to get passed on. Someone may demand something from you using phrases like:

- "We need it immediately."
- "I don't care."
- "That was then, this is now."
- "Figure out how to do it."
- "Make it happen."
- "Deal with it."
- "We can't afford to…"
- "Nothing else matters but…"

It is likely that you will get more than one demand at a time and from different people. Examine the schedule, budget, and headcount available to you. Achieve the request by then scaling down what you would normally do so that it fits in your new triangle. Do the things that will have the most impact on meeting the request to the greatest extent possible.

Unrested

Running long tests after staying up for 30 hours straight or working over 100 hours a week is not the best approach to finding defects. It is, however, a good way to introduce them!

When developers do this, they keep testers in business, but it does not help to get the game released. It is just as bad for the project when testers make mistakes.

Reporting a problem that does not really exist (for example, because you tested the wrong build or did not do the test setup or installation properly) will cause the developers unnecessary re-evaluations and will waste precious time. If you absolutely have to do testing late at night or at the end of a long week, make a checklist to use before and after testing. If there are other testers around, have them check your work, and you can check theirs. Write down relevant information as you go along, so you will not make mistakes later on if you have to rely on your tired memory. If something is wrong, stop what you are doing. Go back and make it right, as it says in the test instructions. After testing is done, record the pertinent results and facts. Here is an example checklist that you can start with and expand upon to fit your own game projects.

Late-Night Testing Checklist

PRE-TEST

Do you have the right version of the test?

Test version: ______________________________

Are you using the right version of the build?

Build version: ______________________________

Are you using the right hardware configuration and settings?

Describe: ______________________________

Are you using the right game controller and settings?

Describe: ______________________________

Which installation options did you use (if any)?

Describe: ______________________________

Is the game in the correct initial state before running the test case?

Describe: ______________________________

POST-TEST

Did you complete all of the steps in order?

Did you record the completion of the tests and the test results?

Did you record all of the problems you found?

If you reported a problem, did you fill in all of the required fields?

In addition to putting formal practices into place for checking mistakes, also look for strategies to prevent them in the first place. Depending on your game platform and test environment, flexible test methods such as Exploratory Testing may be a viable approach for parts of your testing strategy.

"Nearsighted"

Panic symptoms can include too much focus on the near term. Game projects take many months, so make that a factor in deciding what to work on today and how to do it. A good question to ask a tester that will put them in the right frame of mind is "Will this be our last chance to test this?" If the answer is "no," then discuss how to approach the present situation in the context of the overall strategy of repeated testing, feedback from test results, budgeting resources, and so on.

Successful sports teams know how to avoid panic. Even if they are losing, they are confident that they can come back and win the game because they are (a) familiar with the situation, and (b) prepared to deal with it from practice, film study, and in-game experience, (c) rested, and (d) do not feel pressure to make up the deficit immediately. Teams that have a losing record often lack one or more of these ingredients.

RULE TWO: TRUST NO ONE

On the surface, this sounds like a cynical approach, but the very fact that testing is built into the project means that something cannot be trusted. (You read about this in Chapter 4, "How Bugs Happen.") The very existence of testers on a game project is a result of such trust issues as the following:

- The publisher does not trust that your game will release on time and with the promised features, so they write contracts that pay your team incrementally based on demonstrations and milestones.
- The press and public do not trust that your game will be as good and as fun or as exciting as you promised, so they demand to see screenshots and demos, write critiques, and discuss your work on websites and social media.
- Project managers do not trust that the game code can be developed without defects, so testing is planned, funded, and staffed. This can include testers from a third-party QA company or the team's own internal test department.

Do not take any of this lack of trust personally. It is a matter of business, technology, and competition. A great deal of money is on the line, and investors do not want to lose it on your project. The technologies required to produce the game may not even have been available when development started, giving your team the opportunity to create the kinds of games no one has ever done before. By trying to break the game, and failing, you establish confidence that it will work. Games that do not come out right fall victim to rants and complaints for all to see. Do not let this happen to you.

Balancing Act

Evaluate the basis of your testing plans and decisions. Hearsay, opinions, and emotions are elements that can distract you from what you should really be doing. Using test methods and documenting both your work and your results will contribute to an objective game testing environment.

Measuring and analyzing test results, even from past games, will give you data about your game's strengths and weaknesses. The parts that you trust the least, the weak ones, will need the most attention in terms of testing, retesting, and analysis. This relationship is illustrated in Figure 7.4.

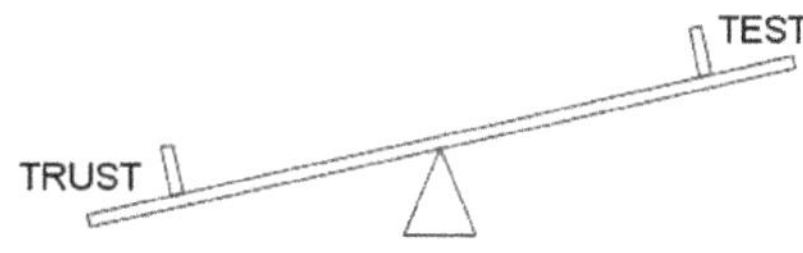

FIGURE 7.4 Low trust means more testing.

The parts you can trust the most, the strong ones, will require the least attention from you, as illustrated in Figure 7.5. These should still be retested from time to time to reestablish your trust.

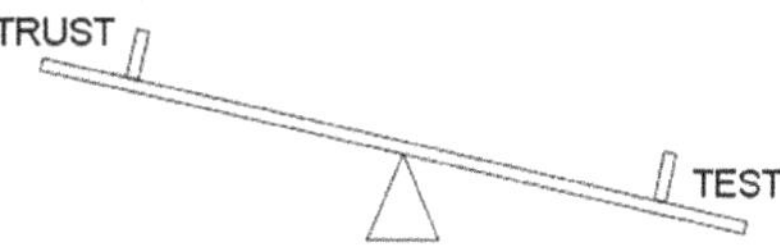

FIGURE 7.5 More trust leads to less testing.

Word Games

It is useful to be wary of advice you get from outside the test team. Well-meaning people will suggest shortcuts so the game development can make better progress, but you will not remove bugs from the game by simply not finding them. Do not trust what these people are telling you. At the same time, do not cross the boundary from being distrustful to turning hostile. The whole team is working to deliver the best game it can, even when it does not seem that way through the eyes of a tester.

A general form of statements to be wary of is "X happened, so (only/don't) do Y." Here are some examples:

- "Only a few lines of code have changed, so don't inspect any other lines."
- "The new audio subsystem works the same as the old one, so you only need to run your old tests."

- "We added foreign language strings for the dialog assets, so just check a few of them in one of the languages and the rest should be okay, too."

Here are some variants of the above:

- "We only made one small change to the character creator, so don't bother testing it."
- "You can run just one or two tests on this and let me know if it works."
- "We have to get this out today, so just..."

Finally, here is the worst statement you can be told:

- "My code doesn't have bugs, so don't waste your time testing it."

Listen to What is Meant, Not What is Said
You may be surprised how many bugs you will find by behaving the opposite from the advice you get about what should and should not be tested.

Do not equate a "trust no one" attitude with a "don't do anything you're asked to do" attitude. If a test lead or the project manager needs you to meet goals for certain kinds of testing to be done, be sure you fulfill your obligation to them before going off and working on the issues you do not trust. The difference is between being a "hero" ("I finished the tests you wanted, and also managed to start looking at the tournament mode and found some problems there. We should do more testing on that next time around.") or a "zero" ("I didn't have time to do the tests you wanted because I was getting some new tests to work for the tournament mode.")

Last Chance

Examine your own tests and look for ways you can improve so you gain more trust in your own skills in finding defects. However, never let that trust turn into arrogance or the belief that you are perfect. Leave room to mistrust yourself just a little bit. Remain open to suggestions from managers, developers, other testers, and yourself. For example, if you are in the middle of running a test and you are not sure you are testing the right version—check it! You may have to go back and start over, but that is better than reporting the wrong results and wasting other people's time in addition to your own.

As the development of a game progresses, management and developers want to feel comfortable about the quality of the game, and, ultimately, its readiness for final release. As a tester, you should not be lulled into complacency. Re-energize your team periodically by instructing them to "Treat this build like it's our last chance to find problems." Conflicts will arise about whether to introduce new tests, and you will hear complaints about why important problems are found so late in the project. There are many reasons for late defects showing up that have nothing to do with incompetent testing. Here are some you might see:

- The defects were introduced late, just as you found them.
- Bugs from earlier rounds of testing blocked you from getting to the part of the game where the late defect was hiding.
- As you spend more time testing the game, you become familiar with where the defects are coming from, so especially subtle problems might not be found until late in the project.

In any case, even if the bugs were there from the very first build, they were not put there by testers.

Trust Fund

You can get a head start on knowing what not to trust in a couple of ways. Sometimes the developers will tell you, if you just ask. Consider the following sample dialogue:

> *Tester: "Hey Alex, is there anything you're particularly concerned about that I should focus on in my testing?"*
>
> *Alex: "Well, we just redid the script for the Fuzzy Sword quest, so we definitely you to look at that."*

You can get more clues by mentioning parts of the system and seeing how people react. Rolling eyes and pauses in response may suggest that there is some doubt as to how well a new weapon or quest will work as well as it did before the latest changes.

MANAGING CRUNCH TIME

At some point in every game project, the developers and testers enter long days and high stress of *crunch time*. During these weeks, people can stay in the office for days at a time, sleep under their desks, eat nothing but carry-out, drink massive amounts of caffeine, and become strangers to their families. All in all, it is an often-unhealthy twilight world where it may seem that the only important thing is finishing the game. (Sledge 2023)

The promise of crunch time is that a team of dedicated people who believe they are working on something special are willing to make sacrifices in other areas of their lives to see the game project creation come out right. The people work hard because they *want* to, because it is important to them, and because it is fun. Their motivation comes from an internal desire rather than an external mandate. If you have ever worked hard in group of people to achieve a cherished goal, you know how exhilarating and rewarding it can be.

However, when it goes poorly, you have people who feel pressured to put in long hours so that they will not lose their jobs, who do not care what is in the game as long it gets done, and who may feel bitter and exploited. If you have ever had to grind away at a pointless task that was doomed to failure anyway, you know how horrible that can be.

When it goes *really* poorly, crunch time turns into a *death march*, which is any period of extraordinary effort that lasts more than one month. Avoid a death march at all costs. The benefits of the team working in what seems like permanent overtime hours are lost in mistakes caused by exhaustion. Apathy sets in. Team morale breaks down. Hours (and eventually days) are spent complaining about delays in the schedule, which lead to even more delays in the schedule. If you ever find yourself saying, "We can make the deadline if everyone works two months of mandatory overtime," take a deep breath, step back, and re-evaluate.

Interview: Madison Cramer

Madison Cramer is currently a Quality Engineer for casino game developer Everi Holdings, but she began her career by working as a tester and test manager on a global "AAA" franchise, where poor communication and semi-permanent crunch time led to burnout, in both herself and her team. "I was working 50- to 70-hour weeks, sometimes without weekends," she said "I had almost nonstop overtime."

Even though she proved to be a talented tester and was quickly promoted to manage the night shift at her studio, she says she found it to be "a fast track to burnout, because doing more in a space where you're already overworking is extremely detrimental to mental health."

Part of what led Ms. Cramer to leave that studio was structural problems she did not find herself in a position to solve. "I feel like it's very easy for QA to be swept under the rug," she says, "or to be considered disposable, or to not be respected." Navigating multiple administrative layers across several locations across the US made the job even harder, because many in upper management "had very little understanding of what our work was like," she says.

Working in casino game development allows her to do the same job, but in a very high-stakes environment. In the gaming sector, a bug in the software might lead to many thousands of dollars of lost revenue if a game pays out too much, or lawsuits and regulatory trouble if it does not pay off as advertised, "So our job is very respected," she says.

In her current position, Cramer says that her quality of life—and her morale—has greatly improved, because she is part of "a very close-knit, really communicative development team where there's just a lot of dialog."

(Permission: Madison Cramer)

Crunch time comes in one form or another to every project. When it arrives, be very careful to not irritate or offend people. As time runs out, emotions run high and tempers can flare. One of the hardest parts of making a game is the last-minute agonizing over how important any given bug is. Such decisions are likely to be made in the supercharged atmosphere of too little time and not enough sleep. In these final days, try to keep your sense of proportion, understand that there is rarely a "right" decision, and remember that even if you disagree with what is happening, you still need to work for the good of the game.

GIVE AND TAKE

If you have been playing close attention up to this point—and you should have, as an aspiring or working game tester—you would have noticed an apparent contradiction between the testing approach to counteract panic (do not treat this build like it is the last one) and the "trust no one" approach of treating each build like it *is* the last one. A sports analogy might illustrate how these concepts can coexist.

In baseball, one batter cannot step up to the plate with bases empty and score six runs. Instead, batter by batter and inning by inning, the team members bat according to the situation, producing the most runs they can. The batters and base runners succeed by being patient, skilled, and committed to their manager's strategy. If every batter tries to hit a home run, the team will strike out often and leave the opposing pitcher fresh for the next inning.

At the same time, when each player is at bat or on base, he is aggressively trying to achieve the best possible outcome. He is fully analyzing the type and location of each pitch, executing his swing properly, and running as fast as he can once the ball is hit. He knows that it contributes to the team's comeback and that one run or RBI could mean the difference between a win or a loss.

So, as a tester, you can do *both at once* by following this advice:

- Know your role on the team based on what responsibilities have been assigned to you.
- Execute your tasks aggressively and accurately.
- Do the most important things first.
- Do the tests most likely to find defects often.
- Make objective and emotion-free decisions as often as you can.

THE REST OF THE STORY

Apply the two rules to what you read in this book. Do not trust that what you read here will work every time for everything that you do. If you get results that do not make sense, find out why. Try a new approach, then evaluate it to decide whether to go on using it or refining it. You may decide to keep trying something new or go back to what you were going to do in the first place but do try the new approach before passing judgment. A word of caution: do not trust yourself too much before you make sure you are applying the technique properly. Then you can decide whether it works for you. You will discover that the methods suggested in this book are sound and worthwhile.

Remember, as a game tester, everyone is trusting in you to find problems before the game ships. Do not give them cause to panic!

EXERCISES

1. Name one common reason for a lack of trust in game development.
2. What are some phrases that might indicate high pressure on a project?
3. If you or your team is tired, what are some ways you can avoid mistakes while you are testing?
4. What is rule one for successful testing?
5. Why should you, as a manager, be concerned with the morale of your team?
6. It is possible to do more testing with less money and fewer people. True or False?

REFERENCE

Sledge, Ben. 2023. "19 Years Later, The EA Spouse Still Has a Point." *The Gamer,* July 22, 2023. *https://www.thegamer.com/ea-spouse-still-has-a-point/*.

CHAPTER 8

COMBINATORIAL TESTING

Test managers and producers are continually struggling with how much testing is too little, too much, or just right. Game quality has to be good enough for players, but testing cannot go on forever if the game is going to make its release date. Trying to test every possible combination of game events, configurations, functions, and options is neither practical nor economical under these circumstances. Taking shortcuts or skipping some testing, however, is risky business.

Pairwise combinatorial testing is a way to find defects and gain confidence in the game software, while keeping the test sets small relative to the amount of functionality they cover. *Pairwise combination* means that each value you use for testing needs to be combined at least once with each other value of the remaining parameters.

PARAMETERS

Parameters are the individual elements of the game that you want to include in your combinatorial tests. You can find test parameters by looking at various types of game elements, functions, and choices such as:

- game events
- game settings
- gameplay options

- hardware configurations
- character attributes
- customization choices

The test you create can be *homogeneous* (designed to test combinations of parameters of the same type) or *heterogeneous* (designed to test more than one type of parameter in the same table).

For example, testing choices from a Game Options screen for their effect on gameplay is done with a homogeneous combinatorial table. If you go through various menus to select different characters, equipment, and options to use for a particular mission, then that results in a heterogeneous table.

VALUES

Values are the individual choices that are possible for each parameter. Values could be entered as a number, entered as text, or chosen from a list. There are many choices for a player to make, but do they all need to be considered in your testing? That is, does every single value or choice have the same weight or probability of revealing a defect, or can you reduce the number of values you test without impacting your test's ability to reveal the defects in the game?

Defaults

Consider whether *default* values should be used in your tests. These are the settings and values that you get if you do not select anything special and just start playing the game as installed. You might also want to consider the first item in any list—say, a choice of hairstyle for your character—to be a kind of default value. Selecting the first item in each of a series of lists allows you to start playing as quickly as possible. Those default choices are the values you will be using.

If the combinatorial testing is the only testing that will be using these parameters, then the defaults should be included. They are the values that will be most often used, so you do not want to let bugs escape that will affect nearly everyone who plays the game.

However, if combinatorial testing is going to be a complement to other types of testing, then you can reduce your test burden by leaving the default values out of your tables. This strategy relies on the fact that the defaults will be used so often that you can expect them to show up in the other testing being done for the game. If you consider leaving these values out, get in touch with the other groups or people who are testing to make sure they do plan on using default values. If you have a test plan for your game, use it to document which sets of tests will incorporate default values and which will not.

Enumerations

Many choices in a game are made from a set of distinct values or options that do not have any particular numerical or sequential relationship to one another. Choosing which car to drive, or which team to play, or which fighter to use are examples of this kind of choice.

Regardless of the number of unique choices (team, car, fighter, weapon, song, hairstyle, and so on), each one should be represented somewhere in your tests. It is easy to find bugs that happen independent of which particular choice is made. Those that do escape tend to happen for only a few of the choices.

Ranges

Many of the game options and choices require the player to pick a number from a range or list. This could be done by directly entering a number or by scrolling through a list to make a selection. For each range of numbers, three particular values tend to have special defect-revealing properties: zero, the minimum, and the maximum.

Any time a zero (0) is presented as a possible choice or entry, it should be included in testing. This is partly due to the unique or ambiguous way that the value 0 might affect the game's source code. Here is a partial list of possible unintended zero-induced effects:

- a loop could prematurely exit or could execute code in the body of the loop before checking for zero
- confusion between starting loop counts at 0 or 1
- confusion with arrays or lists starting at index 0 or 1
- Zero is often used to represent special meaning, such as to indicate an infinite timer or that an error has occurred.

- Zero is the same value as the string termination (NULL) character in C, C++, C#, and Objective-C.
- Zero is the same value as the logical (Boolean) False value in C, C++, C#, and Objective-C.

Minimum values are also a good source of defects. They can be applied to numerical parameters or list choices. Look for the opportunity to use minimum values with parameters related to the following:

- time
- distance
- speed
- quantity
- size
- bet, sell, or purchase amount

For example, using a minimum time might not allow certain effects to be completed once they are started and could make certain goals unachievable.

Maximum values can also cause undesirable side effects. They are especially important to use where they place an extra burden of time or skill for the tester to reach the maximum value. Both developers and testers tend to pass over these values in favor of "easier" testing.

Use maximum values for the same parameter categories that you would test for minimum values. In addition to testing in-game elements, be sure to also include tests for the maximum number of players, maximum number of controllers connected, maximum number of saved files, and maximum storage—disk, cartridge, mobile device memory, and so on.

Boundaries

When a child (or even an adult) colors in a page of a coloring book, we judge how they do by how well they stay within the lines. Likewise, it is the responsibility of the game tester to check the game software around its boundaries. Game behavior that does not "stay within the lines" leads to defects.

Some of the boundaries to test might be physically rendered in the game space, such as the following:

- town, realm, or city borders
- goal lines, sidelines, foul lines, and end lines on a sports field or court
- mission or race waypoints
- start and finish lines
- portal entrances and exits

Other boundaries are not physical. These can include:

- mission, game, or match timers
- the speed that a character or vehicle can achieve
- the distance a projectile can travel
- the distance at which graphic elements become visible, transparent, or invisible

Carefully examine the rules of the game (or the game engine) to identify hidden or implied boundaries.

For example, in football there are rules and activities connected with the timing of the game. The timing of a football game is broken into four quarters of equal length, with a halftime pause in the game that occurs after the end of the second quarter. The game ends if one team has more points than another at the end of the fourth quarter. With two minutes left in each half, the referee stops the clock for the two-minute warning. To test a football game, two-minute quarters are a good boundary value to see if the second and fourth quarters of the game each start normally or with the special two-minute warning. A three-minute duration could also be interesting because it is the smallest duration that would have a period of play prior to the two-minute warning.

Another boundary example is from *Marvel Snap*. When one monthly season comes to an end and a new season begins, the game retains the player's card, card progress, decks, collection level, credits, and gold balance, but resets the player's rank and conquest progress.

CONSTRUCTING TABLES

To see how a combinatorial table is constructed, start with a simple table using parameters that have only two possible values. Games are full of these kinds of parameters, providing choices such as On or Off, Male or Female, Mario or Luigi, or Night or Day. This test combines character attributes for a Jedi character in a Star Wars™ game to test their effects on combat animations and damage calculations. The three test parameters are character gender (in this example, male or female), whether the character uses a one-handed or two-handed lightsaber, and whether the character follows the light side or the dark side of the Force.

The table starts with the first two parameters arranged in the first two columns so that they cover all four possible combinations, as shown in Table 8.1.

TABLE 8.1 First two columns of the Jedi combat test

Gender	Lightsaber
Male	1-handed
Male	2-handed
Female	1-handed
Female	2-handed

To construct a full combinatorial table, repeat each of the Gender and Lightsaber pairs, and then combine each with the two possible Force values. When the "Light" and "Dark" Force choices are added in this way, the size of the table doubles (which is determined by the number of rows, as shown in Table 8.2).

TABLE 8.2 Complete three-way combinatorial table for the Jedi combat test

Gender	Lightsaber	Force
Male	1-handed	Light
Male	1-handed	Dark
Male	2-handed	Light
Male	2-handed	Dark
Female	1-handed	Light
Female	1-handed	Dark
Female	2-handed	Light
Female	2-handed	Dark

For a pairwise combinatorial table, it is necessary to combine each value of every parameter with each value of every other parameter at least once somewhere in the table. A pair that is represented in the table is said to be *satisfied*, while a pair not represented in the table is *unsatisfied*. The following six pairings must be satisfied for the Jedi combat table:

1. Male Gender paired with each Lightsaber choice (1-handed, 2-handed)
2. Female Gender paired with each Lightsaber choice (1-handed, 2-handed)
3. Male Gender paired with each Force choice (Light, Dark)
4. Female Gender paired with each Force choice (Light, Dark)
5. One-handed (1-handed) Lightsaber paired with each Force choice (Light, Dark)
6. Two-handed (2-handed) Lightsaber paired with each Force choice (Light, Dark)

To make the pairwise table, rebuild from Table 8.1 by adding a column for the Force values. Next, enter the light and dark choices for the male character, as shown in Table 8.3. This satisfies pairings 1 and 3—male gender with lightsaber choices and male gender with Force choices.

TABLE 8.3 Adding Force choices for the Male rows

Gender	Lightsaber	Force
Male	1-handed	**Light**
Male	2-handed	**Dark**
Female	1-handed	
Female	2-handed	

Adding the "Dark" value to the first "Female" row will satisfy the criteria for pairing a "1-handed" lightsaber with each Force choice, as illustrated in Table 8.4.

TABLE 8.4 Adding the first Force choice for the Female character test

Gender	Lightsaber	Force
Male	1-handed	Light
Male	2-handed	Dark
Female	**1-handed**	**Dark**
Female	2-handed	

Finally, fill in the light value in the second female row to produce Table 8.5, which completes the pairwise criteria for all parameters. This final entry takes care of the remaining pairings for female gender paired with each lightsaber choice, female gender paired with each Force choice, and two-handed lightsaber paired with each Force choice.

This new table is only half the size of Table 8.2, which was developed to account for all possible three-way combinations, rather than concentrating on using parameter pairs. Including the Force parameter in these tests is "free" in terms of the resulting number of test cases. In many cases, pairwise combinatorial tables let you add complexity and coverage without increasing the number of tests you will need to run. This will not always be true; sometimes you will need a few more tests as you continue to add parameters to the table. The growth of the pairwise table, however, will be much slower than full combinatorial tables for the same set of parameters and their values.

TABLE 8.5 Completed pairwise combinatorial table for the three Jedi combat parameters

Gender	Lightsaber	Force
Male	1-handed	Light
Male	2-handed	Dark
Female	1-handed	Dark
Female	**2-handed**	**Light**

In this simple example, the pairwise technique has cut the number of required tests in half by creating every mathematically possible combination of all of the parameters of interest. This technique and its benefits are not limited to tables with two-value parameters. Parameters with three or more choices can be efficiently combined with other parameters of any dimension. When it makes sense, incorporate more parameters to make your tables more efficient.

The number of choices (values) tested for a given parameter is referred to as its *dimension*. Tables are characterized by the dimensions of each parameter. They can be written in descending order, with a superscript to indicate the number of parameters of each dimension. In this way, the Jedi combat table completed in Table 8.5 is described as a 2^3 table. A table with one parameter of three values, two parameters of four values, and three parameters of two values is described as a $4^2 3^1 2^3$. Another way to describe the characteristics of the table is to list the parameter dimensions individually in descending order with a dash between each value. Using this notation, the Jedi combat table is a 2-2-2 table, and the second example mentioned above is described as 4-4-3-2-2-2. You can see how the second notation takes up a great deal of space when there are a significant number of parameters. Use whichever works best for you.

Create pairwise tables of any size for your game tests using the following short and simple process. These steps might not always produce the optimum (smallest possible) size table, but you will still create an efficient table:

1. Choose the parameter with the highest dimension.
2. Create the first column by listing each test value for the first parameter N times, where N is the dimension of the next highest dimension parameter.

3. Start populating the next column by listing the test values for the next parameter.
4. For each remaining row in the table, enter the parameter value in the new column that provides the greatest number of new pairs with respect to all of the preceding parameters entered in the table. If no such value can be found, alter one of the values previously entered for this column and resume this step.
5. If there are unsatisfied pairs in the table, create new rows and fill in the values necessary to create one of the required pairs. If all pairs are satisfied, then go back to Step 3.
6. Add more unsatisfied pairs using empty spots in the table to create the most possible new pairs. Go back to Step 5.
7. Fill in empty cells with any one of the values for the corresponding column (parameter).

The next example is a little more complicated than the previous one. Use the preceding process to complete a pairwise combinatorial table for some of the *FIFA 15* Match parameters under the Game Settings menu in order to be able to test their effect on the user's visual experience during gameplay. Figure 8.1 shows a portion of the available Match settings. To be thorough, also verify the contents, spelling, capitalization, and punctuation for the description of each setting that appears at the bottom of the Game Settings dialog. For example, when the Difficulty Level selection is highlighted, the description is "Based on your skill, set the difficulty level of your opponent."

Half Length is the real-world amount of time it takes to complete each half of the match. Half Length times are selectable from 4-10 minutes, 15 minutes, or 20 minutes. This test design will use 4 and 10 as range boundaries and 20 because it is the maximum value for this parameter. Match Difficulty Level choices in the game range from "Amateur" to "Legendary," so these two extremes should be represented. Match Referees each have a different level of strictness with regard to calling fouls and issuing cards. We will test using referees that represent a "Lenient," "Average," or "Strict" approach toward fouls and cards. Just be sure that you select a referee that has the same attributes for both card strictness and foul strictness. For example, you can use H. G. Monksfield as the "Lenient" referee, F. Fredskild as the "Average" referee, and M. Barbosa

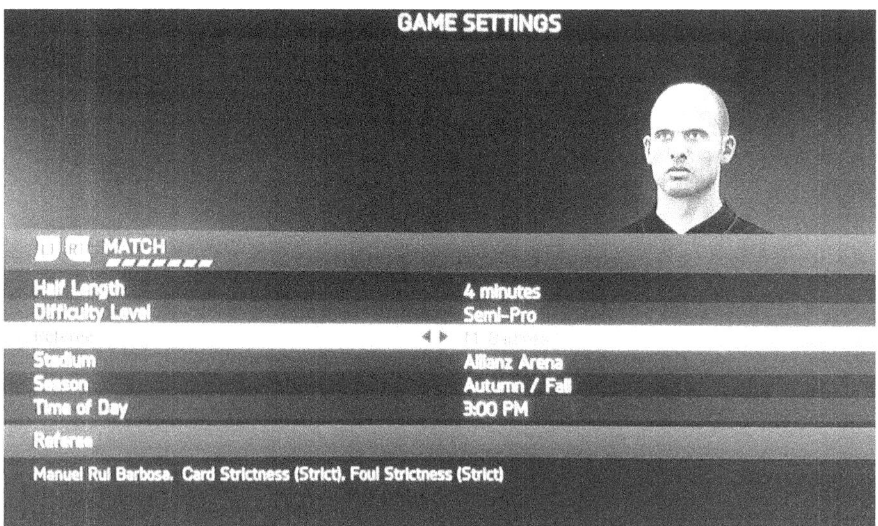

FIGURE 8.1 *FIFA 15* Match settings screen

as the "Strict" referee. Finally, the two extreme Game Speed choices will be tested: "Slow" and "Fast." As a result, you will create a $3^3 2^3$ table consisting of three parameters with three choices (Half Length, Referee, and Weather), followed by three parameters with two choices (Difficulty Level, Pitch Wear, and Game Speed).

It does not matter if you are not familiar with the game or the detailed rules of soccer. You just need to be able to understand and follow the seven steps for constructing a pairwise combinatorial table.

Begin the process with Steps 1 and 2, and list the Half Length values three times in the first column of the table. This is because Half Length is one of the parameters with the highest dimension (3). One of the parameters with the next highest dimension is Referee, which also has a dimension of three.

Next, apply Step 3 and put each of the three Referee values in the first three rows of column 2. Table 8.6 shows what the matrix looks like at this point. A row number is included in the table so that each combination (test case) can be referenced easily.

TABLE 8.6 Starting the *FIFA 15* Match settings text table

	Half Length	**Referee**
1	4 min	Lenient
2	10 min	Average
3	20 min	Strict
4	4 min	
5	10 min	
6	20 min	
7	4 min	
8	10 min	
9	20 min	

Apply Step 4 to continue filling in the next column. Starting with the fourth row, enter a Referee parameter that creates the highest number of new pairs. Because this is only the second column, you can create only one new pair. The "Lenient" Referee has already been paired with "4 min" Half Length, so you can put "Average" in row 4 to create a new pair. Likewise, "Strict" and "Lenient" should go in rows 5 and 6 to create new pairs with "10 min" and "20 min," respectively. Table 8.7 shows the resulting combinations at this point in the process.

TABLE 8.7 Adding the second set of Referee values

	Half Length	**Referee**
1	4 min	Lenient
2	10 min	Average
3	20 min	Strict
4	4 min	Average
5	10 min	Strict
6	20 min	Lenient
7	4 min	
8	10 min	
9	20 min	

Continue with Step 4 to complete the Referee column. At the seventh row, enter a Referee type that creates a new pair with the "4 min" Half Length value. "Lenient" (row 1) and "Average" (row 4) have already been paired, so "Strict" is the correct value for this row. By the same process, "Lenient" goes in row 8 and "Average" in row 9. Table 8.8 shows the first two columns completed in this manner.

TABLE 8.8 Completing the Referee column

	Half Length	Referee
1	4 min	Lenient
2	10 min	Average
3	20 min	Strict
4	4 min	Average
5	10 min	Strict
6	20 min	Lenient
7	4 min	Strict
8	10 min	Lenient
9	20 min	Average

Applying Step 5, check that all of the pairs required for the first two columns are satisfied:

- Half Length = "4 min" is paired with "Lenient" (row 1), "Average" (row 4), and "Strict" (row 7).
- Half Length = "10 min" is paired with "Lenient" (row 8), "Average" (row 2), and "Strict" (row 5).
- Half Length = "20 min" is paired with "Lenient" (row 6), "Average" (row 9), and "Strict" (row 3).

Because all of the pairs required for the first two columns are represented in the table, Step 5 sends us back to Step 3 to continue the process with the Weather option and its three test values. Applying Step 3, list the "Clear," "Rainy," and "Overcast" Weather values at the top of the third column, as shown in Table 8.9.

TABLE 8.9 Starting the Weather column

Row	Half Length	Referee	Weather
1	4 min	Lenient	Clear
2	10 min	Average	Rainy
3	20 min	Strict	Overcast
4	4 min	Average	
5	10 min	Strict	
6	20 min	Lenient	
7	4 min	Strict	
8	10 min	Lenient	
9	20 min	Average	

Proceed with Step 4 to add the Weather value that creates the most pairs for row 4 ("4 min" and "Average"). "Clear" is already paired with "4 min" in row 1, and "Rainy" is already paired with "Average" in row 2, so "Overcast" is the correct entry for this row. In the same manner, "Clear" creates two new pairs in row 5, and "Rainy" creates two new pairs in row 6. Table 8.10 shows what the test table looks like at this point.

TABLE 8.10 Adding the second set of Weather values

Row	Half Length	Referee	Weather
1	4 min	Lenient	Clear
2	10 min	Average	Rainy
3	20 min	Strict	Overcast
4	**4 min**	**Average**	**Overcast**
5	**10 min**	**Strict**	**Clear**
6	**20 min**	**Lenient**	**Rainy**
7	4 min	Strict	
8	10 min	Lenient	
9	20 min	Average	

Again, continue with Step 4 to complete the Weather column. "Rainy" produces two new pairs in row 7: "Overcast" in row 8, and "Clear" in row 9. Table 8.11 shows the completed Weather column.

TABLE 8.11 Completing the Weather column

Row	Half Length	Referee	Weather
1	4 min	Lenient	Clear
2	10 min	Average	Rainy
3	20 min	Strict	Overcast
4	4 min	Average	Overcast
5	10 min	Strict	Clear
6	20 min	Lenient	Rainy
7	**4 min**	**Strict**	**Rainy**
8	**10 min**	**Lenient**	**Overcast**
9	**20 min**	**Average**	**Clear**

It is time again to check that all the required pairs are satisfied. Because the first two columns have been previously verified, there is no need to check them again. Check the new Weather column against all of its predecessors, as follows:

- Half Length = "4 min" is paired with "Clear" (row 1), "Rainy" (row 7), and "Overcast" (row 4).
- Half Length = "10 min" is paired with "Clear" (row 5), "Rainy" (row 2), and "Overcast" (row 8).
- Half Length = "20 min" is paired with "Clear" (row 9), "Rainy" (row 6), and "Overcast" (row 3).
- Referee = "Lenient" is paired with "Clear" (row 1), "Rainy" (row 6), and "Overcast" (row 8).
- Referee = "Average" is paired with "Clear" (row 9), "Rainy" (row 2), and "Overcast" (row 4).
- Referee = "Strict" is paired with "Clear" (row 5), "Rainy" (row 7), and "Overcast" (row 3).

With all of the required pairs satisfied at this point, Step 5 sends you back to Step 3 to add the Difficulty parameter. Table 8.12 shows the two Difficulty test values added to the top of the fourth column.

TABLE 8.12 Starting the Difficulty choices

Row	Half Length	Referee	Weather	Difficulty
1	4 min	Lenient	Clear	Beginner
2	10 min	Average	Rainy	Legendary
3	20 min	Strict	Overcast	
4	4 min	Average	Overcast	
5	10 min	Strict	Clear	
6	20 min	Lenient	Rainy	
7	4 min	Strict	Rainy	
8	10 min	Lenient	Overcast	
9	20 min	Average	Clear	

Apply Step 4, and add the extreme Difficulty values in row 3 ("20 min," "Strict" and "Legendary") that create the most pairs with column 4. Either "Beginner" or "Legendary" will create a new pair with all three of the other values in this row. For this exercise, choose "Beginner" for row 3. Continue from there, and add the correct values for rows 4 through 6. "Beginner" in row 4 would create only one new pair with "Beginner," so "Legendary" is the right value to put here, creating pairs with "4 min" and "Overcast." Rows 5 and 6 are populated with "Legendary" to create two new pairs in each of these rows as well: "Strict" and "Clear" in row 5 and "20 min" and "Rainy" in row 6. Table 8.13 shows the table with the newly satisfied pairs in **bold** in rows 3 through 6.

TABLE 8.13 Generating new Difficulty pairs

Row	Half Length	Referee	Weather	Difficulty
1	4 min	Lenient	Clear	Beginner
2	10 min	Average	Rainy	Legendary
3	**20 min**	**Strict**	**Overcast**	**Beginner**
4	**4 min**	Average	**Overcast**	**Legendary**
5	10 min	**Strict**	**Clear**	**Legendary**
6	**20 min**	**Lenient**	Rainy	**Legendary**
7	4 min	Strict	Rainy	
8	10 min	Lenient	Overcast	
9	20 min	Average	Clear	

Now choose the right Difficulty values for the remaining rows. "Legendary" in row 7 does not create any new pairs, because "4 min" is already paired with "Legendary" in row 4, "Strict" is already paired with "Legendary" in row 5, and "Rainy" is already paired with "Legendary" in row 6. "Beginner" in row 7 *does* create a new pair with "Rainy," so it is the only correct choice. Rows 8 and 9 must be populated with "Beginner" to create new pairs with Half Length = "10 min" and Referee = "Average." Table 8.14 shows the completed Difficulty column.

TABLE 8.14 Completing the Difficulty column

Row	Half Length	Referee	Weather	Difficulty
1	4 min	Lenient	Clear	Beginner
2	10 min	Average	Rainy	Legendary
3	20 min	Strict	Overcast	Beginner
4	4 min	Average	Overcast	Legendary
5	10 min	Strict	Clear	Legendary
6	20 min	Lenient	Rainy	Legendary
7	4 min	Strict	**Rainy**	**Beginner**
8	**10 min**	Lenient	Overcast	**Beginner**
9	20 min	**Average**	Clear	**Beginner**

Now check that all the required pairs for the Difficulty column are satisfied:

- Half Length = "4 min" is paired with "Beginner" (rows 1, 7) and "Legendary" (row 4).
- Half Length = "10 min" is paired with "Beginner" (row 8) and "Legendary" (rows 2, 5).
- Half Length = "20 min" is paired with "Beginner" (rows 3, 9) and "Legendary" (row 6).
- Referee = "Lenient" is paired with "Beginner" (rows 1, 8) and "Legendary" (row 6).
- Referee = "Average" is paired with "Beginner" (row 9) and "Legendary" (rows 2, 4).
- Referee = "Strict" is paired with "Beginner" (rows 3, 7) and "Legendary" (row 5).
- Weather = "Clear" is paired with "Beginner" (rows 1, 9) and "Legendary" (row 5).
- Weather = "Rainy" is paired with "Beginner" (row 7) and "Legendary" (rows 2, 6).
- Weather = "Overcast" is paired with "Beginner" (rows 3, 8) and "Legendary" (row 4).

Having satisfied all of the pairs required by the Difficulty column, go back again to Step 3 to continue with the Pitch Wear option. Add the "Beginner" and "Legendary" Difficulty values to the top of the fifth column, as shown in Table 8.15.

TABLE 8.15 Starting the Pitch Wear column

Row	Half Length	Referee	Weather	Difficulty	Pitch Wear
1	4 min	Lenient	Clear	Beginner	None
2	10 min	Average	Rainy	Legendary	Heavy
3	20 min	Strict	Overcast	Beginner	
4	4 min	Average	Overcast	Legendary	
5	10 min	Strict	Clear	Legendary	
6	20 min	Lenient	Rainy	Legendary	
7	4 min	Strict	Rainy	Beginner	
8	10 min	Lenient	Overcast	Beginner	
9	20 min	Average	Clear	Beginner	

Step 4 requires a value in column 5 that creates the most value for row 3. Only "Heavy" creates a new pair with the four other values in this row. Repeat for row 4, and choose "None," which creates three new pairs ("Average," "Overcast," and "Legendary"), while "Heavy" would provide only one new pair with "4 min." Populating rows 5 and 6 with "None" creates two new pairs in each of these rows, while "Heavy" would add only one new pair in each case. Table 8.16 shows "Heavy" chosen for "Scrolling Lineups" in row 3, and "None" for rows 4, 5, and 6.

TABLE 8.16 Adding to the Pitch Wear column

Row	Half Length	Referee	Weather	Difficulty	Pitch Wear
1	4 min	Lenient	Clear	Beginner	None
2	10 min	Average	Rainy	Legendary	Heavy
3	20 min	Strict	Overcast	Beginner	Heavy
4	4 min	Average	Overcast	Legendary	None
5	10 min	Strict	Clear	Legendary	None
6	20 min	Lenient	Rainy	Legendary	None
7	4 min	Strict	Rainy	Beginner	
8	10 min	Lenient	Overcast	Beginner	
9	20 min	Average	Clear	Beginner	

A "Heavy" value in the remaining rows produces a new pair for each: "4 min," "Lenient," and "Clear." Table 8.17 shows the completed Pitch Wear column.

TABLE 8.17 Completing the Pitch Wear column

Row	Half Length	Referee	Weather	Difficulty	Pitch Wear
1	4 min	Lenient	Clear	Beginner	None
2	10 min	Average	Rainy	Legendary	Heavy
3	20 min	Strict	Overcast	Beginner	Heavy
4	4 min	Average	Overcast	Legendary	None
5	10 min	Strict	Clear	Legendary	None
6	20 min	Lenient	Rainy	Legendary	None
7	**4 min**	Strict	Rainy	Beginner	**Heavy**
8	10 min	**Lenient**	Overcast	Beginner	**Heavy**
9	20 min	Average	**Clear**	Beginner	**Heavy**

It is time again to check that all the required pairs for the new column are satisfied:

- Half Length = "4 min" is paired with "Beginner" (rows 1, 4) and "Heavy" (row 7).
- Half Length = "10 min" is paired with "Beginner" (row 5) and "Heavy" (rows 2, 8).
- Half Length = "20 min" is paired with "None" (row 6) and "Heavy" (rows 3, 9).
- Referee = "Lenient" is paired with "None" (rows 1, 6) and "High" (row 8).
- Referee = "Average" is paired with "None" (row 4) and "Heavy" (rows 2, 9).
- Referee = "Strict" is paired with "None" (row 5) and "Heavy" (rows 3, 7).
- Weather = "Clear" is paired with "None" (rows 1, 5) and "Heavy" (row 9).
- Weather = "Rainy" is paired with "None" (row 6) and "Heavy" (rows 2, 7).
- Weather = "Overcast" is paired with "None" (row 4) and "Heavy" (rows 3, 8).

- Difficulty = "Beginner" is paired with "None" (row 1) and "Heavy" (rows 3, 7, 8, 9).
- Difficulty = "Legendary" is paired with "None" (rows 4, 5, 6) and "Heavy" (row 2).

This confirms that the pairs required for the Difficulty column are all satisfied. The process sends you back to Step 3 to pair the Game Speed values in the final column. Add the "Slow" and "Fast" values to the top of this column, as shown in Table 8.18.

TABLE 8.18 Starting the Game Speed column

Row	Half Length	Referee	Weather	Difficulty	Pitch Wear	Game Speed
1	4 min	Lenient	Clear	Beginner	None	Slow
2	10 min	Average	Rainy	Legendary	Heavy	Fast
3	20 min	Strict	Overcast	Beginner	Heavy	
4	4 min	Average	Overcast	Legendary	None	
5	10 min	Strict	Clear	Legendary	None	
6	20 min	Lenient	Rainy	Legendary	None	
7	4 min	Strict	Rainy	Beginner	Heavy	
8	10 min	Lenient	Overcast	Beginner	Heavy	
9	20 min	Average	Clear	Beginner	Heavy	

As you proceed from here, something new happens. Either of the Game Speed values added to row 3 creates four new pairs, so neither value can be selected. A "Slow" creates new pairs with "20 min," "Strict," "Overcast," and "Heavy," while a "Fast" creates new pairs with "20 min," "Strict," "Overcast," and "Beginner." As you go through the table, you will find that no preferred value can be found for any of the remaining rows. Do not trust us on this (remember Rule Two?)—check for yourself! According to Step 4, "If no such value can be found, alter one of the values previously entered for this column and resume this step." So, one of the Game Speed values in the first two rows should be changed. Table 8.19 shows the updated table with the second Game Speed value changed to "Slow."

TABLE 8.19 Restarting the Game Speed column

Row	Half Length	Referee	Weather	Difficulty	Pitch Wear	Game Speed
1	4 min	Lenient	Clear	Beginner	None	Slow
2	10 min	Average	Rainy	Legendary	Heavy	**Slow**
3	20 min	Strict	Overcast	Beginner	Heavy	
4	4 min	Average	Overcast	Legendary	None	
5	10 min	Strict	Clear	Legendary	None	
6	20 min	Lenient	Rainy	Legendary	None	
7	4 min	Strict	Rainy	Beginner	Heavy	
8	10 min	Lenient	Overcast	Beginner	Heavy	
9	20 min	Average	Clear	Beginner	Heavy	

Continue to Step 4 from this point and see that there are now clear choices for the remaining rows. A "Fast" in row 3 provides new pairs with all of the first five columns, versus only four new pairs that would be provided by a "Slow." Another "Fast" in row 4 provides four new pairs versus three from using "Slow," and rows 5 and 6 get two new pairs from a "Fast" versus only one from a "Slow." Table 8.20 shows how the table looks with these values filled in.

TABLE 8.20 Adding to the Game Speed column

Row	Half Length	Referee	Weather	Difficulty	Pitch Wear	Game Speed
1	4 min	Lenient	Clear	Beginner	None	Slow
2	10 min	Average	Rainy	Legendary	Heavy	Slow
3	**20 min**	**Strict**	**Overcast**	**Beginner**	**Heavy**	**Fast**
4	**4 min**	**Average**	Overcast	**Legendary**	**None**	**Fast**
5	**10 min**	Strict	**Clear**	Legendary	None	**Fast**
6	20 min	**Lenient**	**Rainy**	Legendary	None	**Fast**
7	4 min	Strict	Rainy	Beginner	Heavy	
8	10 min	Lenient	Overcast	Beginner	Heavy	
9	20 min	Average	Clear	Beginner	Heavy	

Complete the final three values for the table in the same manner. "Slow" is the only value that produces a new pair in each of these rows. The completed table is shown in Table 8.21.

TABLE 8.21 The completed Match Game Settings test table

Row	Half Length	Referee	Weather	Difficulty	Pitch Wear	Game Speed
1	4 min	Lenient	Clear	Beginner	None	Slow
2	10 min	Average	Rainy	Legendary	Heavy	Slow
3	20 min	Strict	Overcast	Beginner	Heavy	Fast
4	4 min	Average	Overcast	Legendary	None	Fast
5	10 min	Strict	Clear	Legendary	None	Fast
6	20 min	Lenient	Rainy	Legendary	None	Fast
7	4 min	**Strict**	Rainy	Beginner	Heavy	**Slow**
8	10 min	Lenient	**Over-cast**	Beginner	Heavy	**Slow**
9	**20 min**	Average	Clear	Beginner	Heavy	**Slow**

Now, for perhaps the last time, check that all the required pairs for the Game Speed column are satisfied:

- Half Length = "4 min" is paired with "Slow" (rows 1, 7) and "Fast" (row 4).
- Half Length = "10 min" is paired with "Slow" (row 2) and "Fast" (rows 5, 8).
- Half Length = "20 min" is paired with "Slow" (row 9) and "Fast" (rows 3, 6).
- Referee = "Lenient" is paired with "Slow" (rows 1, 8) and "Fast" (row 6).
- Referee = "Average" is paired with "Slow" (rows 2, 9) and "Fast" (row 4).
- Referee = "Strict" is paired with "Slow" (row 7) and "Fast" (rows 3, 5).
- Weather = "Clear" is paired with "Slow" (rows 1, 9) and "Fast" (row 5).
- Weather = "Rainy" is paired with "Slow" (rows 2, 7) and "Fast" (row 6).
- Weather = "Overcast" is paired with "Slow" (row 8) and "Fast" (rows 3, 4).

- Difficulty = "Beginner" is paired with "Slow" (rows 1, 7, 8, 9) and "Fast" (row 3).
- Difficulty = "Legendary" is paired with "Slow" (row 2) and "Fast" (rows 4, 5, 6).
- Pitch Wear = "None" is paired with "Slow" (row 1) and "Fast" (rows 4, 5, 6).
- Pitch Wear = "Heavy" is paired with "Slow" (rows 2, 7, 8, 9) and "Fast" (row 3).

Well done! By creating a pairwise combinatorial table, you developed nine test cases that can be used to test these game parameters and values comprising 216 possible mathematical combinations (3*3*3*2*2*2). It was certainly worth the effort to create the table in order to save 207 test cases! Also note that for this table you did not have to resort to Steps 6 and 7. That will not be true in every case, so do not rule it out for the future.

Now you are ready to test the game using the combinations in the table, and check for any irregularities or discrepancies with what you expect to happen. Create test tables as early as possible, for example, by using information provided in the design document long before any working code is produced. Check any available documentation to see if there is a clear definition of what should happen for each of your combinations. That will equip you to raise questions about the game that perhaps had not been considered. This is an easy way to prevent bugs and improve gameplay.

A second approach is to ask people involved with code or requirements "What happens if…" and read your combinations. You might be surprised by how many times you will get an answer like "I don't know" or "I'll have to check and get back with you." This is a much more economical alternative to finding surprises late in the project. It is also much more likely that your issues will be fixed, or at least considered, by the time the code is written.

Do not just check for immediate or near-term effects of your combinatorial tests. It is important to make sure that a menu selection is available or a button performs its function when pressed, but mid-term and far-term effects can lock up or spoil the game down the road. Some of these effects to consider are as follows:

- Does my game or session end properly?
- Do achievements get recorded properly?

- Can I progress to appropriate parts of the game or story?
- Did actions taken in the game get properly counted toward season/career accomplishments and records?
- Can I properly start and play a new session?
- Can I store and retrieve sessions or files?

Take a look at the section "Issues with Goal Scoring," which is a first-person account in which the tester observed unusual behaviors by running a similar set of combinatorial tests for *FIFA 11* Match settings.

"Issues with Goal Scoring"

In the first long game I tested, I noticed late into the game that when play resumed after scoring a goal, the time of the goal and the goal scorer was being reported incorrectly. The time was in the past and the same time and player's name was always being shown. Examining the game's Match Events screen showed that the goal which was being reported was the 30th goal that was scored by my team. Going through a few other test cases from the table confirmed that this happened consistently and occurred whether the 30th goal is scored in the first half or second half of the game.

A second issue appeared when I finished one match with a score of 107-0. The individual Player Ratings: Goals screen recorded a total of only 100 goals and the Match Events: Goals screen listed only the 100 most recent goals—goals scored in the first 4 minutes were not listed.

In the games I played with the 10-minute and 20-minute Half Length, I experienced some delays and stuttering on screen transitions during gameplay toward the end of the match. Perhaps this was due to the accumulation of all of the events that get logged during the game, a graphics memory management issue, or some other nefarious problem.

At some point in the sequence of games I was playing, I noticed that when starting a game with Weather = Rainy, the rain is initially visible in the "lobby," but stops coming down after a short period of time. There was also no rain coming down during the match. This is one of those cases where the tester needs to run further experiments to determine which combination or sequence of combinations triggers this phenomenon.

Figure 8.2 shows how a second-half goal scored in the 55th minute of the match is reported as occurring in the 23rd minute of the previous half.

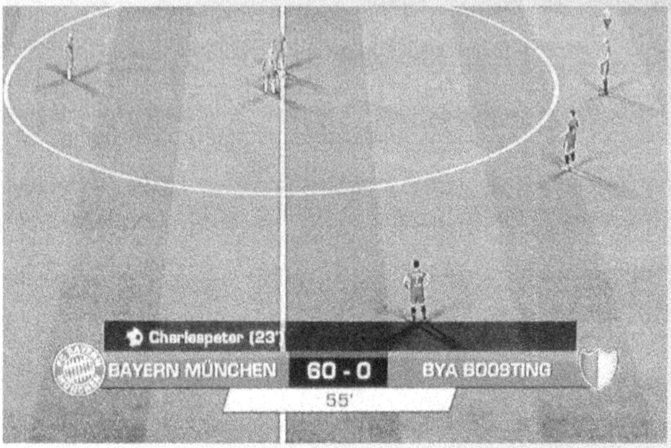

FIGURE 8.2 *FIFA 11:* Incorrect reporting of the goal event.

COMBINATORIAL TEMPLATES

Some pre-constructed tables are included in Appendix D, with the companion files for this book. You can use them by substituting the names and values of the parameters you want to test for the entries in the template. This will be a fast way to produce tables of fewer than 10 tests without having to develop them from scratch and then verify that all of the necessary pairs are covered. Wherever a "*" appears after a letter in the template, such as "B*," that means you can substitute any of the test values for that parameter and the table will still be correct.

NOTE *All companion files for this book may be obtained by contacting the publisher at info@merclearning.com.*

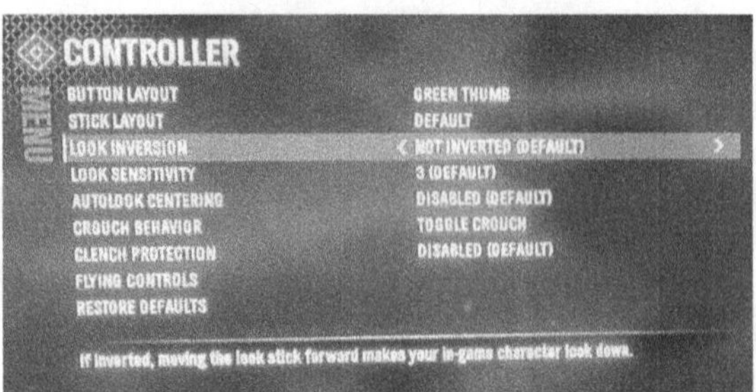

FIGURE 8.3 Advanced Controller settings for *Halo: Reach*

To see how this works, create a test table based on the Advanced Controls settings screen for *Halo: Reach* (Figure 8.3). Start by determining how many parameters and values you want to test. Figure 8.3 shows nine Advanced Controls parameters and examples of their values. For this exercise, test the Look Inversion, Look Sensitivity, AutoLook Centering, Crouch Behavior, and Clench Protection settings in combination with one another. The Look Sensitivity parameter can be a value from 1 to 10 and the remaining parameters have Yes/No, Enabled/Disabled, or Toggle/Hold choices. Because Look Sensitivity ranges from 1 to 10, a good set of values to test would be the default, minimum, and maximum values, which are 3, 1, and 10, respectively. This test requires a combinatorial table of five parameters, where one parameter value (Look Sensitivity) has three values, and the remaining parameters have two test values. Scan through Appendix D, and you will find that Table D.18 corresponds to this configuration.

For each parameter, assign one of the test values to the alphanumeric placeholders in the table template. Because Look Sensitivity is the only parameter with three values, it goes in the first column. The default value (3) will be assigned to A1, the minimum value (1) to A2, and the maximum (10) to A3. Replace each instance of A1, A2, and A3 in the table with their assigned values. The table at this point should look like Table 8.22.

TABLE 8.22 Sensitivity values placed into the table template

Test	Look Sensitivity	Param B	Param C	Param D	Param E
1	**3**	B1	C1	D1	E1
2	**1**	B2	C2	D1	E1
3	**10**	B1	C2	D2	E1
4	**3**	B2	C2	D2	E2
5	**1**	B1	C1	D2	E2
6	**10**	B2	C1	D1	E2

Next, choose one of the two-value parameters, and substitute its name and values in the template's ParamB column. Choose the Look Inversion parameter, assigning the default value "NO" to each instance of "B1" in the table and the "YES" value to each "B2." The table now looks like Table 8.23.

TABLE 8.23 Look Inversion values added to the table

Test	Look Sensitivity	Look Inversion	Param C	Param D	Param E
1	3	**NO**	C1	D1	E1
2	1	**YES**	C2	D1	E1
3	10	**NO**	C2	D2	E1
4	3	**YES**	C2	D2	E2
5	1	**NO**	C1	D2	E2
6	10	**YES**	C1	D1	E2

Continue this process for the remaining columns using the default values for the first entry and the remaining value for the other choice. The complete design is shown in Table 8.24.

TABLE 8.24 Completed Controller settings table

Test	Look Sensitivity	Look Inversion	AutoLook Centering	Crouch Behavior	Clench Protection
1	3	NO	NO	HOLD	DISABLED
2	1	YES	YES	HOLD	DISABLED
3	10	NO	YES	TOGGLE	DISABLED
4	3	YES	YES	TOGGLE	ENABLED
5	1	NO	NO	TOGGLE	ENABLED
6	10	YES	NO	HOLD	ENABLED

To use one of the template files included in the book's companion files, start by selecting the appropriate file based on your table dimensions. If all of your test parameters have only two values, then use the file CombTemplates2Values.xls. If one or more of your parameters has three values, use the file CombTemplates3Values.xls. If you have any parameters with four or more values, then you need to construct your table by hand or see the "Combinatorial Tools" section that follows.

Once you have identified the right template file to use, click the tab at the bottom of the worksheet that corresponds to the number of test parameters you are using. Then find the template on that sheet that matches your parameter configuration.

For the *Halo: Reach* Controller settings test, you would open the CombTemplates3Values.xls file, and click the "5 params" tab at the bottom of the worksheet. Scroll down until you find the table labeled "1 parameter with 3 values, 4 parameters with 2 values." You will see that this table is identical in structure to the one in Appendix C that produced the test in Table 8.24. Cut this table out and paste it into your own test file. Finally, do a textual substitution for each of the test values to arrive at the same result.

COMBINATORIAL TEST GENERATION

At some point, you will find it difficult to construct and verify large parameter and value counts. Fortunately, James Bach has made a tool, available to the public at *https://www.satisfice.com/download/allpairs*, which handles this for you. (It is also among the companion files available for this book.) The Allpairs tool uses a tab-delimited text file as input and produces an output file that includes a pairwise combinatorial table as well as a report on how many times each pair was satisfied in the table.

To use Allpairs, start by creating a file that contains tab-delimited columns of parameter names with the test values in the following table. Here is an example based on the match settings from the fighting game *Dead or Alive 3* (DOA3):

TABLE 8.25 The match settings from *Dead or Alive 3*

Difficulty	MatchPoint	LifeGauge	RoundTime
Normal	1	Smallest	NoLimit
Easy	2	Small	30
Hard	3	Normal	40
VeryHard	4	Large	50
	5	Largest	60
			99

Remember, this is not an attempt at a combinatorial table; the Allpairs tool will provide that. This is a description of the parameters you want to test: Difficulty, MatchPoint, LifeGauge, and RoundTime. Even though there are only four parameters, the fact that they have 4, 5, 5, and 6 values each to test would make this difficult to construct and validate by hand. That also means there are 600 (4*5*5*6) values if you try to test all 4-way combinations. You should expect a much smaller test set from a pairwise combinatorial test of these options—somewhere in the 30 to 40 range—based on the dimensions of the two largest parameters (6*5).

Now open a Command Prompt window and enter `allpairs input.txt > output.txt`, where `input.txt` is the name of your tab-delimited parameter list file, and `output.txt` is the name of the file where you want to store the generated combinatorial table. Make sure you are in the directory where the files are located, or provide the full path.

For this *DOA3* table, the command might be `allpairs doa-params.txt > doapairs.txt`. Here is what the test case portion of the output looks like:

TEST CASES

TABLE 8.26 A portion of the *Dead or Alive 3* test case output

Case	Difficulty	MatchPoint	LifeGauge	RoundTime	Pairings
1	Normal	1	Smallest	NoLimit	6
2	Easy	2	Small	NoLimit	6
3	Hard	3	Normal	NoLimit	6
4	VeryHard	4	Large	NoLimit	6
5	Hard	1	Small	30	6
6	VeryHard	2	Smallest	30	6
7	Normal	3	Large	30	6
8	Easy	4	Normal	30	6
9	VeryHard	1	Normal	40	6
10	Hard	2	Large	40	6
11	Easy	3	Smallest	40	6
12	Normal	4	Small	40	6

(Continued)

TABLE 8.26 A portion of the *Dead or Alive 3* test case output (continued)

Case	Difficulty	MatchPoint	LifeGauge	RoundTime	Pairings
13	Easy	1	Large	50	6
14	Normal	2	Normal	50	6
15	VeryHard	3	Small	50	6
16	Hard	4	Smallest	50	6
17	Normal	5	Largest	60	6
18	Easy	1	Largest	60	4
19	Hard	2	Largest	60	4
20	VeryHard	3	Largest	60	4
21	Easy	5	Smallest	99	5
22	Normal	4	Largest	99	4
23	Hard	5	Small	99	4
24	VeryHard	5	Normal	99	4
25	~Normal	5	Large	NoLimit	2
26	~Easy	5	Largest	30	2
27	~Hard	5	Largest	40	2
28	~Very-Hard	5	Largest	50	2
29	~Hard	4	Smallest	60	2
30	~Hard	1	Large	99	2
31	~Very-Hard	~1	Largest	NoLimit	1
32	~Normal	~1	Small	60	1
33	~Easy	~2	Normal	60	1
34	~Easy	~3	Large	60	1
35	~Normal	2	~Smallest	99	1
36	~Easy	3	~Small	99	1

Imagine how much time you would have spentdoing that by hand! The Case and Pairings columns are added to the output by the Allpairs tool. The Case value is a sequential number uniquely identifying each test case. The Pairings number indicates how many necessary parameter pairs are represented by the set of values in each row. For example, the Pairings value in row 18 is "4." You can check for yourself that row 18 produces four new pairs: "Easy-Largest," "Easy-60," "1-Largest," and "1-60." The "Largest-60" pair was satisfied earlier in the table at row 17, and the "Easy-1" pair first appears in row 13.

Values that begin with the "~" symbol are wildcards. That is, any value of that parameter could be placed there without removing one of the necessary pairings to complete the table. The tool arbitrarily chooses, but you, the knowledgeable tester, can replace those with more common or notorious values, such as defaults or values that have caused defects in the past.

The output from Allpairs also produces a Pairing Details list, which is an exhaustive list of each necessary pair and all of the rows that include that pair. One of the pairings listed for the *DOA3* table is

MatchPoint Difficulty 1 Easy 13,18

which means that the pair "MatchPoint = 1" and "Difficulty = Easy" occurs 2 times, in rows 13 and 18 of the table.

In the same list, the entry

RoundTime LifeGauge 60 Largest 4 17,18,19,20

traces the "RoundTime = 60" and "LifeGauge = Largest" pair to rows 17–20 of the combinatorial table. This kind of information is especially useful if you want to limit your testing to all the instances of a particular pair. One reason for doing that would be to limit verification testing of a release that fixed a bug caused by one specific pair.

Another use for the Pairing Details information is to quickly narrow down the possible cause of a new defect by immediately testing the other entries in the table that had the same pairs as the test that just failed. For example, if the test in row 13 fails, search the Pairing Details list for other pairs that were included in row 13. Then run the tests on any rows listed in addition to row 13. Here are the pairs that are satisfied by row 13:

TABLE 8.27 *Dead or Alive 3:* sample pairs satisfied by row 13

RoundTime	MatchPoint	50	1	1	13
RoundTime	LifeGauge	50	Large	1	13
RoundTime	Difficulty	50	Easy	1	13
MatchPoint	LifeGauge	1	Large	2	13, 30
MatchPoint	Difficulty	1	Easy	2	13, 18
LifeGauge	Difficulty	Large	Easy	2	13, 34

From this information, tests 18, 30, and 34 could be run next to help identify the pair that causes the defect. If none of those tests fail, then the cause is narrowed down to the first three pairs, which are only found in row 13: "50-1," "50-Large," or "50-Easy." If test 18 fails, then look for the "1-Easy" pair to be the cause of the problem. Likewise, if test 30 fails, then suspect the "1-Large" combination. If test 34 fails, you can suggest "Large-Easy" as the cause of the problem in your defect report.

The Allpairs output file is tab-delimited so you can paste it right into Excel or any other program supporting that format.

COMBINATORIAL ECONOMICS

The examples used in this chapter have produced tables with significant efficiency, covering hundreds of potential combinations in no more than a few dozen tests. As it turns out, these are very modest examples. Some configurations can yield reductions of more than 100:1, 1000:1, and even beyond 1,000,000:1. It all depends on how many parameters you use and how many test values you specify for each parameter. Do you always want to do *less* testing?

Some game features are so important that they deserve more thorough testing than others. One way to use pairwise combinatorial tests for your game is to do full combinatorial testing for critical features and pairwise for the rest. Suppose you identify 10% of your game features as "critical" and that each of these features has an average of 100 tests associated with them (approximately a 4 × 4 × 3 × 2 matrix). It is reasonable to expect that the remaining 90% of the features could be tested using pairwise combinatorial tables and only cost 20 tests per feature. The cost of full combinatorial testing of all features is 100*N, where N is the total number of features to be tested. The cost of pairwise combinatorial testing 90% of those features is 100*0.1*N + 20*0.9*N = 10*N+18*N = 28*N. This provides a 72% savings by using pairwise for the noncritical 90%.

Another way to use combinatorial tests in your overall strategy is to create some tables to use as "sanity" tests. The number of tests you run early in the project will stay low, and then you can rely on other ways of doing "traditional" or "full" testing once the game can pass the sanity tests. Knowing which combinations work properly can also help you select which scenarios to feature in pre-release videos, walkthroughs, or public demos.

In each of these situations, the least expensive way for your team to find and remove the defects is to create pairwise combinatorial tables as early as possible in the development cycle and to investigate the potential results of each test case. Once a design specification becomes available, create combinatorial tables based on the information available to you at the time and question the designers about the scenarios you generate.

If you know your testing budget in terms of staff, effort, or dollars early in the project, you have to make choices about how to distribute your resources to test the game the best you can. Pairwise combinatorial tests provide a good balance of breadth and depth of coverage, which allows you to test more areas of the game than if you concentrate resources on just a few areas.

EXERCISES

1. Explain the difference between a pairwise combinatorial table and a full combinatorial table.
2. Explain the difference between a parameter and a value.
3. Use the appropriate template to add the Offsides (On/Off) parameter from the Game Settings: Rules screen to the *FIFA 15* Match settings test table in Table 8.21.
4. Because some of the issues found with *FIFA 11* are related to the Half Length, add three new rows to the *FIFA 15* Match settings test table that pair the "15 min" Half Length with the other five parameters.
5. Use the Allpairs tool to create a combinatorial table for some of the settings for the mobile game *Kingturn RPG* available on iOS and Android. The first parameter to test is Sound using the values "On" and "Off." The second parameter is Difficulty with the values "Casual," "Normal," "Strategist," "Master," and "King." For the third parameter, use Perma Knockout with values "On" and "Off." Finally, include the Pinch Zoom values for "Slowest," "Slower," "Default," "Faster," and "Fastest."

CHAPTER 9

TEST FLOW DIAGRAMS

Test Flow Diagrams (TFDs) are graphical models representing game behaviors from the player's perspective. Testing takes place by traveling through the diagram to exercise the game in both familiar and unexpected ways.

TFDs provide a formal approach to test design that promotes modularity and completeness. Testers can enjoy a high degree of TFD reuse if the same behaviors are consistent across multiple game titles or features. This benefit extends to sequels and ports to other platforms. The graphical nature of the TFD gives testers, developers, and project managers the ability to easily review, analyze, and provide feedback on test designs.

TFD ELEMENTS

A TFD is created by assembling various drawing components called *elements*. These elements are drawn, labeled, and interconnected according to certain rules. Following the rules will make it possible for your tests to be understood throughout your test organization and makes them easier to reuse in future game projects. The rules will become even more important if your team develops software tools to process or analyze the TFD contents.

Flows are drawn as a line connecting one game "state" to another, with an arrow indicating the direction of flow. Each flow also has a unique

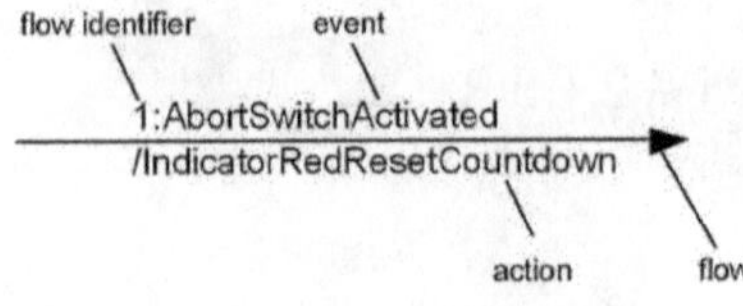

FIGURE 9.1 Flow components

identification number, one *event* and one *action*. A colon (":") separates the event name from the flow ID number and a slash ("/") separates the action from the event. During testing, you *do* what is specified by the event and then *check* for the behavior specified by both the action and the flow's destination state. An example flows and each of its components are shown in Figure 9.1.

Events

Events are operations initiated by the user, peripherals, multiplayer networks, or internal game mechanisms. Think of an event as something that is explicitly done during the game. Picking up an item, selecting a spell to cast, sending a chat message to another player, and an expiring game timer are all examples of events. The TFD does not have to represent all possible events for the portion of the game being tested. It is left up to each tester, who is now in the role of a test designer, to use his knowledge and judgment in selecting the right events that will achieve the purpose of a single TFD or a set of related TFDs. There are three factors that should be considered for including a new event:

1. possible interactions with other events
2. unique or important behaviors associated with the event
3. unique or important game states that are a consequence of the event

Only one event can be specified on a flow, but multiple operations can be represented by a single event. An event name can appear multiple times on a TFD when each instance carries the exact same meaning. Events could possibly cause a transition to a new game state.

Actions

An *action* exhibits temporary or transitional behavior in response to an event. It is something for the tester to check as a result of causing or performing an event. Actions can be perceived through human senses and gaming platform facilities, including sounds, visual effects, game controller feedback, and information sent over a multiplayer game network. Actions do not persist over time. They can be perceived, detected, or measured when they occur but can no longer be perceived, detected, or measured sometime later.

Only one action can be specified on a flow, but multiple operations can be represented by a single action. An action name can appear multiple times on a TFD when each instance carries the exact same meaning.

States

States represent persistent game behavior and are re-entrant. As long as you do not exit the state, you will continue to observe the same behavior and each time you return to the state you should detect the exact same behavior.

A state is drawn as a "bubble" with a unique name inside. If the same behavior applies to more than one state on your diagram, consider whether they could be the same state. If so, remove the duplicates and reconnect the flows accordingly. Each state must have at least one flow entering and one flow exiting.

Primitives

Events, actions, and states are also referred to as *primitives*. Primitive definitions provide details of the behavior represented on the TFD without cluttering the diagram. Primitive definitions form a *Data Dictionary* for the TFD. These definitions could be in text (English), a software language (C++), or an executable simulation or test language (TTCN). See the Data Dictionary section below for details and examples.

Terminators

Terminators are special boxes placed on the TFD that indicate where testing starts and where it ends. Exactly two terminators should appear on each TFD. One is the "IN" box, which normally has a single flow that goes to a state. The other is the "OUT" box, which has one or more flows entering from one or more states.

TFD DESIGN ACTIVITIES

Creating a TFD is not just a matter of mechanically typing or drawing some information you already have in another form. It is a design activity which requires the tester to become a *designer*. A sound approach to getting your TFDs off and running is to go through three stages of activities: Preparation, Allocation, and Construction.

Preparation

Collect sources of game feature requirements.

Identify the requirements that fall within the scope of the planned testing, based on your individual project assignment or the game's Test Plan. This would include any storyboards, design documents, demo screens, or formal software requirements, as well as legacy titles that the new game is based on.

Allocation

Estimate the number of TFDs required and map game elements to each.

Separate large sets of requirements into smaller chunks and try to cover related requirements in the same design. One way to approach this is to test various *abilities* provided in the game, such as picking up a weapon, firing a weapon, healing, and so forth. Plan on having one or more TFDs for each ability, depending on how many variations exist, such as distinct weapon types or different ways to regain health. Another approach is to map situations or scenarios to individual TFDs with a focus on specific *achievements*. These could be individual missions, quests, matches, or challenges, depending on the type of game you are testing. In this case, you are establishing that particular goals or outcomes are achievable according to which path you take in the game. A TFD design based on achievements could be used either instead of, or in addition to, the abilities approach. Do not try to include too much information in a single design. It is easier to complete and manage a few simple TFDs than one that is complex.

Construction

Model game elements on their assigned TFDs using a "player's perspective."

A TFD should not be based on any actual software design structures within the game. The TFD is meant to represent the tester's interpretation of what she expects to happen as the game flows to and from the game states represented on the diagram. Creating a TFD is not as mechanical as constructing a combinatorial table. There is an element of art to it. TFDs for the same game feature could turn out quite differently depending on which tester developed them.

Begin the TFD with a blank sheet or a template. You can start on paper and then transfer your work to an electronic form or do the whole thing at one time on your computer. The use of templates is discussed later in this chapter. Follow the steps below to begin constructing your TFD from scratch. An example appearing later in this chapter illustrates the application of these steps:

1. Open a file and give it a unique name that describes the scope of the TFD.
2. Draw a box near the top of the page and add the text "IN" inside of it.
3. Draw a circle and put the name of your first state inside of it.
4. Draw a flow going from the "IN" box to your first state. Add the event name "Enter" to the flow.

NOTE *Do not number any of the flows at this time. This will be done at the end to avoid record keeping and editing the numbers if you change the diagram during the rest of the design process.*

Unlike the steps given for developing a pairwise combinatorial table in the prior chapter, "Combinatorial Testing," the middle steps for creating a Test Flow Diagram do not have to be followed in any particular order. Construct your diagram as your mind flows through the game scenario you are testing. The creation of the diagram should be iterative and dynamic as the diagram itself raises questions about possible events and their outcomes. Refer to the steps below when you have a problem or when you think you are done to make sure you do not leave out any parts of the process:

1. From your first state, continue to add flows and states. Flows can be connected back to the originating state to test required behavior that is transient (action) or missing (ignored, resulting in no action).
2. Record the traceability of each flow to one or more requirements, options, settings, or functions. This could be as simple as ticking it off from a list or highlighting portions of the Game Design Document, or it can be done formally by documenting this information in a Requirements Traceability Matrix (RTMX).
3. For each flow going from one state (A) to another state (B), check the requirements for possible ways to go from B to A, and add flows

as appropriate. If the requirements neither prohibit nor allow the possibility, review this with the game, feature, or level designer to determine if a requirement is missing (most likely), wrong, or ambiguous. Once all requirements are traced to at least one flow, check the diagram for alternative or additional ways to exercise each requirement. If a flow seems appropriate, necessary, or obvious but cannot be traced to any game documentation, determine if there might be a missing or ambiguous requirement. Otherwise, consider whether the flow is outside of the defined scope of the TFD currently being constructed. Go through these final steps in the order they appear here:

4. Add the "OUT" box.
5. Select which state or states should be connected to the "OUT" box. Your criteria should include choosing places in the test that are appropriate for stopping one test and starting the next one or selecting states that naturally occur at the end of the *ability* or *achievement* modeled by the TFD. For each of these states, provide a connecting flow to the "OUT" box with an "Exit" event. There should be no more than one such flow coming from any state.
6. Update your "IN" and "OUT" box names to "IN_xxx" and "OUT_xxx," where "xxx" is a brief descriptive name for the TFD. This is done at the end in case your scope or focus has changed during the process of creating the TFD.
7. Number all of the flows.

A TFD EXAMPLE

To draw a TFD, you need a drawing application that can draw circles, lines with arrows, rounded or square rectangles, and the ability to attach numbers and text to each element. PowerPoint is an adequate and quite accessible tool for this work, or you might prefer the richer features in Visio or SmartDraw.

Your first TFD example will be based on the ability to pick up a weapon and its ammo while the game properly keeps track of your ammo count and performs the correct audible and visual effects. This is an ability required in first-person shooters, role playing games, action/adventure games, arcade games, and even some racing games.

NOTE

Use your favorite drawing tool to create your own diagram files as you follow the examples in this chapter. Do your own layout and editing and then compare what you designed with the example diagrams each step along the way.

All TFDs start with an "IN" box, followed by a flow to the first state of the game that you want to observe or that you need to reach in order to begin testing. Do not begin every test with the startup screen unless that is what you are trying to test with the TFD. Go to the point in the game where you want to start doing things (events) with the game that you want the tester to check (actions, states).

In this TFD, the first state will represent the situation where the player has no weapon and no ammo. Draw a flow to connect the "IN" box to the "NoGunNoAmmo" state. Per the process described earlier in this chapter, provide the event name "Enter" on the flow, but do not provide an ID number yet. Figure 9.2 shows how the TFD looks at this point.

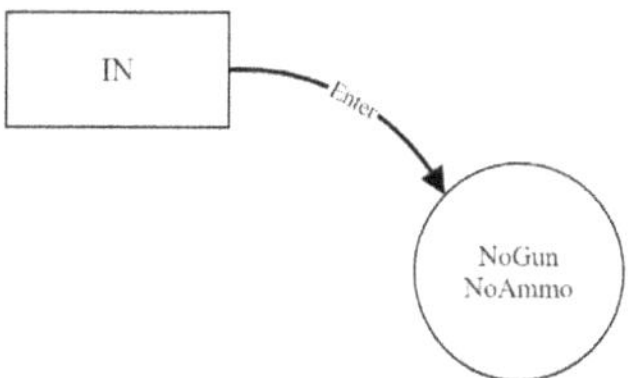

FIGURE 9.2 Starting the "Ammo" TFD

The next step is to model what happens when the player does something in this situation. One likely response is to find a gun and pick it up. Having a gun creates observable differences from not having a gun. A gun appears in your inventory, your character is shown holding the gun, and a crosshair now appears at the center of the screen. These are reasons to create a separate state for this situation. Keep the naming simple and call the new state "HaveGun." In the process of getting the gun, the game could produce some temporary effects like playing the sound of a weapon being picked up and identifying the weapon on the display. The temporary effects are represented by an action on the flow. Name the flow's event "GetGun," and name the action "GunEffects." The TFD with the gun flow and new state is shown in Figure 9.3.

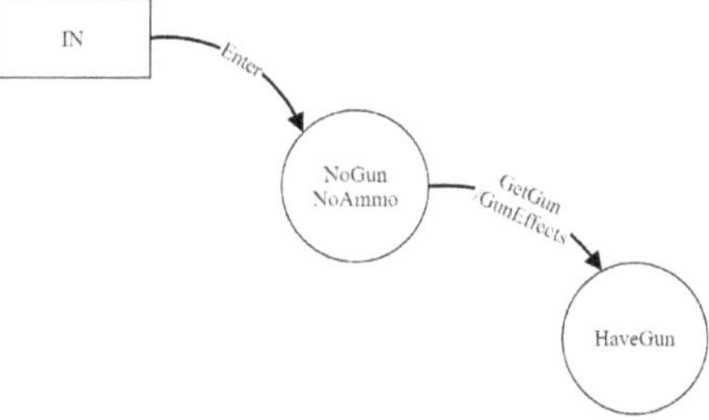

FIGURE 9.3 TFD after picking up a weapon

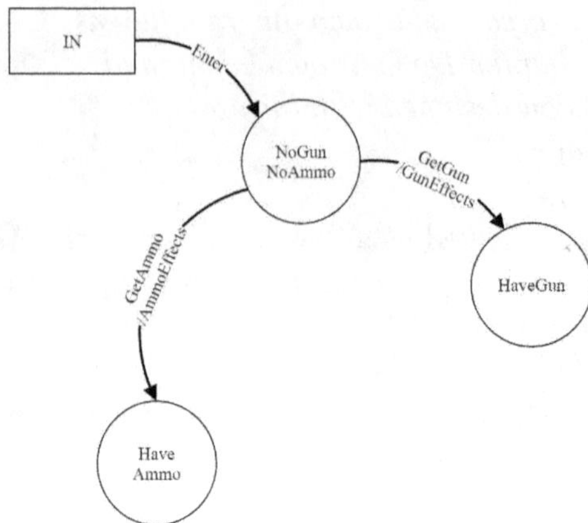

FIGURE 9.4 TFD with "HaveGun" and "HaveAmmo" states

Because it is possible that the player could find and pick up ammo before getting the weapon, add another flow from "NoGunNo-Ammo" to get ammo and check for the ammo sound and visual effects. A new destination state should also be added. Call it "HaveAmmo" to be consistent with the "HaveGun" state name format. Your TFD should look like Figure 9.4 at this point.

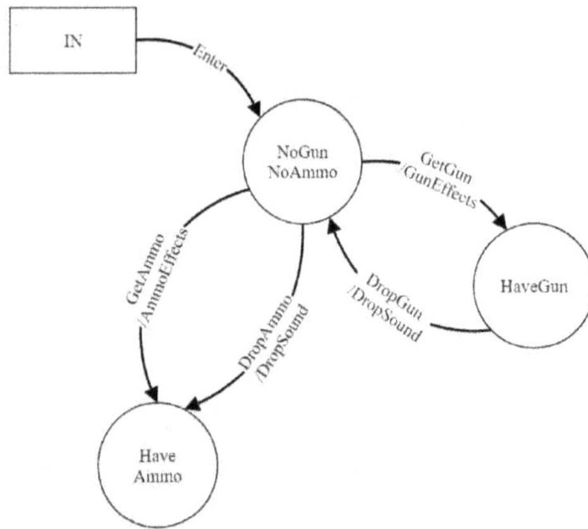

FIGURE 9.5 Return flows added from "HaveGun" and "HaveAmmo"

Now that there are a few states on the diagram, check if there are any flows you can add that go back from each state to a previous one. You got to the "HaveGun" state by picking up a weapon. It could also be possible to go back to the "NoGunNoAmmo" state by dropping the weapon. Likewise, there should be a flow from "HaveAmmo" going back to "NoGunNoAmmo" when the player somehow drops his ammo. If there are multiple ways to do this, each should appear on your TFD. One way might be to remove the ammo from your inventory and another might be to perform a reload function. For this example, just add the generic "DropAmmo" event and its companion "DropSound" action. To illustrate how actions might be reused within a TFD, the diagram will reflect that the same sound is played for dropping either a weapon or ammo. That means the "DropGun" event will also cause the "DropSound" action. The return flows from "HaveGun" and "HaveAmmo" are shown in Figure 9.5.

Now that the test represents gun-only and ammo-only states, connect the two concepts by grabbing ammo once you have the gun. Call the resulting state "HaveGun-HaveAmmo." You should recognize that picking up the gun once you have the ammo will also take you to this very same state. Figure 9.6 shows the new flows and the "HaveGunHaveAmmo" state added to the TFD.

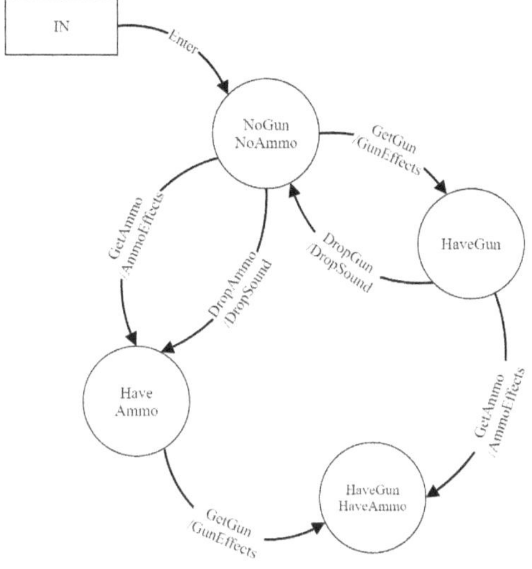

FIGURE 9.6 Flows added to get both gun and ammo

You perhaps have noticed that when new states are added, it is good to leave some room on the diagram for flows or states that you might decide to add when you get further into the design process. Use up some of that empty space now by doing the same thing for "HaveGunHaveAmmo" that you did with the "HaveAmmo" and "HaveGun" states: create return flows to represent what happens when the gun or the ammo is dropped. One question that arises is whether the ammo stays in your inventory or is lost when the gun is dropped. This test is based on the ammo automatically loading when you have the matching weapon, so the "DropGun" event will take you all the way from "HaveGunHaveAmmo" to "NoGunNoAmmo." Be careful not to get caught up in the symmetry that sometimes arises from the diagram. Flows coming out of states do not always return to the previous state. The TFD with these additional flows is shown in Figure 9.7.

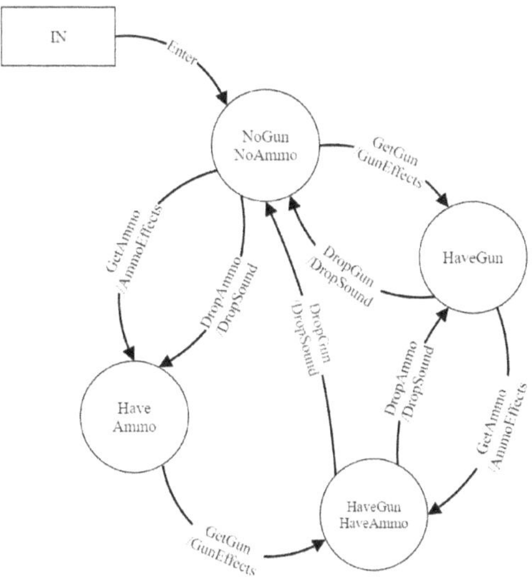

FIGURE 9.7 Return flows added from "HaveGunHaveAmmo"

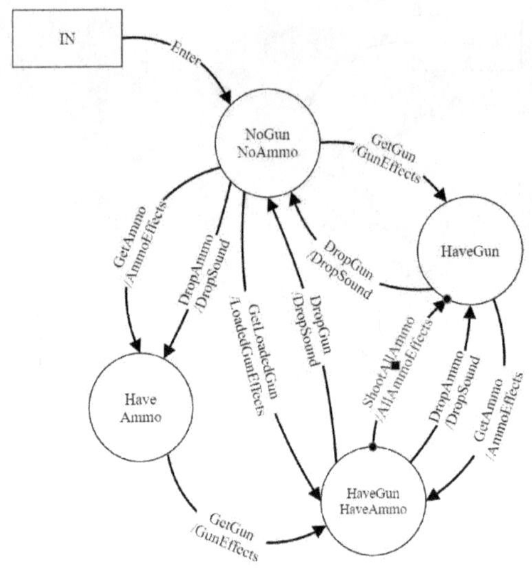

FIGURE 9.8 Loaded gun and shooting flows added

At this point, evaluate whether there is anything else that could be added that remains consistent with the purpose of this test. That is, are there any ways to manipulate the ammo or the gun that would require new flows or states on the TFD? Start from the furthest downstream state and work your way up. If you have the gun and ammo, is there any other way to end up with the gun and no ammo besides dropping the ammo? Well, shooting the gun uses ammo, so you could keep shooting until all of the ammo is used up and then end up back at “HaveGun.” Because both of the states involved in this transition are already on the diagram, you need to add a new flow only from “HaveGun-HaveAmmo” to “HaveGun.” Likewise, besides picking up an empty gun, you might get lucky and get one with some ammo in it. This creates a new flow from “NoGunNoAmmo” to “HaveGunHaveAmmo.” Figure 9.8 shows the diagram with these new interesting flows added.

Note that some of the existing flows were moved around slightly to make room for the new flows and their text. “ShootAllAmmo” will cause sounds, graphic effects, and damage to another player or the environment. Doing “GetLoadedGun” will cause effects similar to the combined effects of separately picking up an unloaded gun and its ammo. The actions for these new events were named “AllAmmoEffects” and “LoadedGunEffects” to reflect the fact that these multiple effects are supposed to happen and need to be checked by the tester. The “ShootAllAmmo” event illustrates that your test events do not have to be atomic. You do not need a separate event and flow for firing each individual round of ammo, unless that is exactly what your test is focusing on.

Do the same for “HaveGun” and “HaveAmmo” that you just did for “HaveGunHaveAmmo.” Question whether there are other things that could happen in those states to cause a transition or a new kind of action. You should recognize that you can attempt to fire the weapon at any time, whether or not you have ammo, so a flow should come out from

"HaveGun" to represent the game behavior when you try to shoot with no ammo. Where does this flow go to? It ends up right back at "HaveGun." This is drawn as a loop as shown in Figure 9.9.

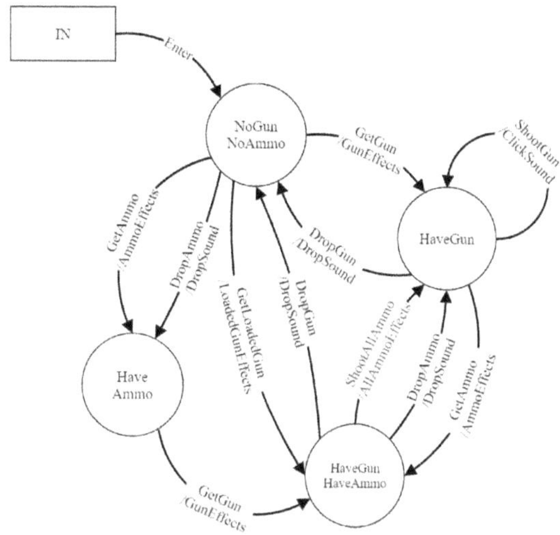

FIGURE 9.9 Flow added to shoot gun with no ammo

At this point, only two things remain to do according to the procedures given earlier in this chapter: add the "OUT" box and number the flows. Keep in mind that the numbering is totally arbitrary. The only requirement is that each flow has a unique number.

Another thing that has been done is to name the "IN" and "OUT" boxes to identify this specific TFD which might be part of a collection of multiple TFDs created for various features of a game. This also makes it possible to uniquely specify the test setup and tear-down procedures in the Data Dictionary definition for these boxes. This is described in further detail later in this chapter.

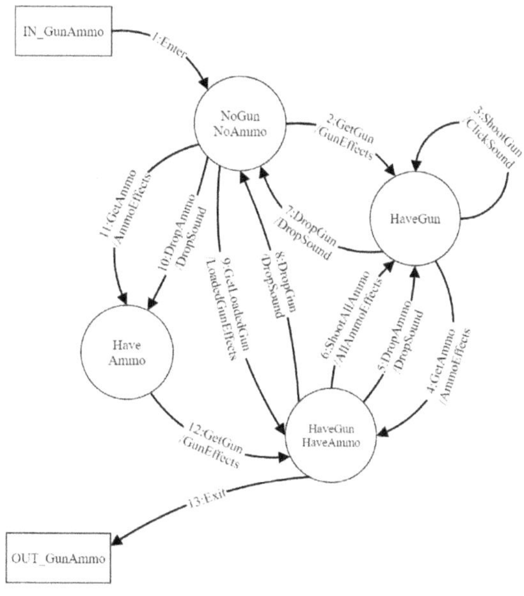

FIGURE 9.10 The completed "Ammo" TFD

Once you complete your diagram, be sure to save your file and give it an appropriate descriptive name. Figure 9.10 shows the completed "Ammo" TFD.

DATA DICTIONARY

The Data Dictionary provides detailed descriptions for each of the uniquely named primitive elements in your TFD collection. This also implies that any primitive name you reuse within a TFD and across multiple TFDs will carry the same meaning during testing. Think of the primitive names on the TFD as a hyperlink to pages that contain their definitions. When you mentally "click" on one of those names, you get the same definition, regardless of which instance of the name you click on.

Data Dictionary Application

If you use SmartDraw to create and maintain your TFDs, you can do this by highlighting the text for an event, action, or state and selecting "Insert Hyperlink" from the "Tools" pulldown menu. Then, manually browse for a text or HTML file that contains the description of the primitive. If you use HTML files for the description, then you can also export your diagram to make your test accessible as a Web page. Do this by selecting "Publish to the Web" from the "File" menu.

It is up to you to decide how formal your definitions should be. In small teams intimate with the product, the TFD by itself might be sufficient if you can trust the person running the test (Rule Two) to remember and consistently apply all of the details of each primitive. For large teams, especially when new people are moving in and out of the test team during the course of the project, the Data Dictionary will provide more consistent and thorough checking, as well as better adherence to the intent of the test. You will perhaps also want to keep TFD use informal in early development stages until the development team better understands how they really want the game to behave. Once the game stabilizes, capture that information in the Data Dictionary.

Data Dictionary Reuse

The Data Dictionary can also be an important tool for reusing your TFDs for different games or game elements. For example, the "Ammo" TFD in Figure 9.10 refers abstractly to "Gun" and "Ammo." Most games involving weapons provide multiple types of weapons and ammo that is specific for each. You could cover this by making copies of the TFD for each of the different weapon types, changing the event, action, and state names to match. An alternative is to keep a generic TFD and then apply different

Data Dictionaries to interpret the TFD specifically for each weapon and ammo type.

A good strategy for *Unreal Tournament* or any other first-person shooter game would be to use a single TFD but have different data dictionaries for the various weapon/ammo pairs such as Flak Cannon/Flak Shells, Rocket Launcher/Rocket Pack, Shock Rifle/Shock Core, and so on. Each Data Dictionary could elaborate on the different audio, visual, and damage effects associated with each pair.

Data Dictionary Example

Build the Data Dictionary by defining each of the elements in the diagram. The "do" items (events) are written normally. The "check" items (actions and states) should be written in list form with a leading dash or bullet to visually separate them from the "do" items. You can also use an empty box character q that can be checked off as the test is run. This is useful for providing a physical record of what the tester observed.

Some of the "Ammo" TFD data dictionary items for Figure 9.10 are defined below for the Bio-Rifle weapon, arranged in alphabetical order for easy searching.

AmmoEffects

❑ *Check that the Bio-Rifle ammo sound is made.*

❑ *Check that the game temporarily displays "You picked up some Bio-Rifle ammo" in white text above the gun icons at the bottom of the screen.*

❑ *Check that the temporary text on the display fades out slowly.*

DropGun

❑ *Hit the "\" key to drop your selected weapon.*

DropSound

❑ *Check that the item drop sound is made.*

Enter

❑ *Selected a match and click the FIRE button to start the match.*

Exit

- ❑ *Hit the ESC key and exit the match.*

GetAmmo

- ❑ *Find a Bio-Rifle ammo pack on the floor in the arena and walk over it.*

GetGun

- ❑ *Find an unloaded Bio-Rifle hovering above the floor of the arena and walk into it.*

GetLoadedGun

- ❑ *Find a Bio-Rifle loaded with ammo hovering above the floor of the arena and walk into it.*

GunEffects

- ❑ *Check that the Bio-Rifle sound is made.*
- ❑ *Check that the game temporarily displays "You got the Bio-Rifle" in white text above the gun icons at the bottom of the screen.*
- ❑ *Check that the game simultaneously displays "Bio-Rifle" temporarily in blue text above the "You got the Bio-Rifle" message.*
- ❑ *Check that all temporary text on the display fades out slowly.*

HaveAmmo

- ❑ *Check that the Bio-Rifle icon is empty in the graphical weapon inventory at the bottom of the screen.*
- ❑ *Check that the Bio-Rifle barrel is not rendered in front of your character.*
- ❑ *Check that you cannot select the Bio-Rifle weapon using the mouse wheel.*
- ❑ *Check that the aiming reticle in the center of the screen has not changed.*

HaveGun.

- ❑ *Check that the Bio-Rifle icon is present in the graphical weapon inventory at the bottom of the screen.*
- ❑ *Check that the Bio-Rifle barrel is rendered in front of your character.*
- ❑ *Check that you can select the Bio-Rifle weapon using the mouse wheel.*
- ❑ *Check that the Bio-Rifle aiming reticle appears as a small blue broken triangle in the center of the screen.*
- ❑ *Check that the ammunition count in the right-hand corner of the screen is 0.*

HaveGunHaveAmmo

- ❑ *Check that the Bio-Rifle icon is present in the graphical weapon inventory. at the bottom of the screen.*
- ❑ *Check that the Bio-Rifle barrel is rendered in front of your character.*
- ❑ *Check that you can select the Bio-Rifle weapon using the mouse wheel.*
- ❑ *Check that the Bio-Rifle aiming reticle appears as a small blue broken triangle in the center of the screen.*
- ❑ *Check that the ammunition count in the right-hand corner of the screen is 40.*

IN_GunAmmo

Launch Unreal Tournament *on the test PC.*

LoadedGunEffects

- ❑ *Check that the Bio-Rifle sound is made.*
- ❑ *Check that the game temporarily displays "You got the Bio-Rifle" in white text above the gun icons at the bottom of the screen.*
- ❑ *Check that the game simultaneously displays "Bio-Rifle" temporarily in blue text above the "You got the Bio-Rifle" message.*

❑ *Check that all temporary text on the display fades out slowly.*

NoGunNoAmmo

❑ *Check that the Bio-Rifle icon is empty in the graphical weapon inventory at the bottom of the screen.*

❑ *Check that the Bio-Rifle barrel is not rendered in front of your character.*

❑ *Check that you cannot select the Bio-Rifle weapon using the mouse wheel.*

FIGURE 9.11 *Unreal Tournament 2004* Bio-Rifle "AmmoEffects"

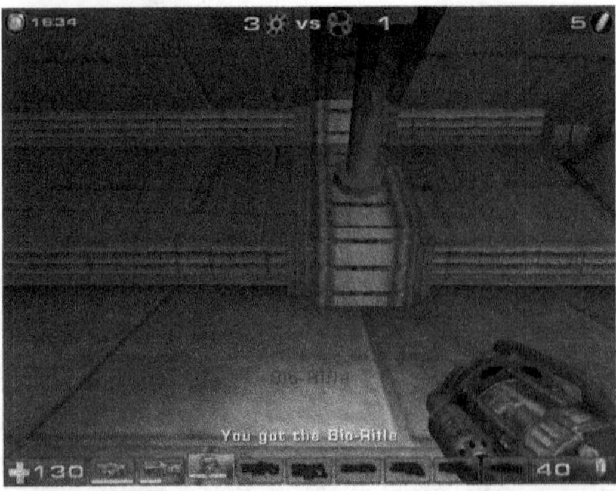

FIGURE 9.12 *Unreal Tournament 2004* Bio-Rifle "GunEffects"

OUT_GunAmmo

At the main menu, click on "EXIT" to exit the game.

You can even include screenshots, art from design documents, or art from storyboards to provide a visual reference for the tester. This works well with the hyperlink and Web publishing approach. The reference graphics can be updated to reflect changes and maturing of the screen layout and art as the game gets closer to completion. For testing the Bio-Rifle, the "AmmoEffects" definition could include a screenshot like the one in Figure 9.11, which shows how the "picked up" confirmation text is rendered on the screen.

Likewise, Figure 9.12 illustrates a useful reference

for showing the Bio-Rifle "GunEffects" action by capturing the on-screen indications that the Bio-Rifle has been picked up and is now the player's active weapon.

TFD PATHS

A test path is a series of flows, specified by the flow numbers in the sequence in which they are to be traversed. Paths begin at the "IN" state and end at the "OUT" state. A set of paths provides behavior scenarios appropriate for prototyping, simulation, or testing.

A *path* defines an individual test case which can be "executed" to explore the game's behavior. Path execution follows the events, actions, and states on the TFD. A textual script can be constructed by cutting and pasting primitives in the order they occur along the path. Testers then follow the script to execute each test, referring to the Data Dictionary for details of each primitive. Automated scripts are created in the same manner, except lines of code are being pasted together rather than textual instructions for a human tester.

Many paths are possible for a single TFD. Tests can be executed according to single strategy for the duration of the project or path sets can vary according to the maturity of the game code as it progresses through different milestones. The TFD remains constant as long as the correct game requirements and behaviors do not change. Some useful strategies for selecting test paths are described below.

Minimum Path Generation

This strategy is designed to produce the smallest number of paths that will result in covering all of the flows in the diagram. In this context, "covering" means that a flow is used at least once somewhere in the test.

The benefits of using a minimum path set are that you have a low test count and the knowledge that you exercised all parts of the diagram at least once. The drawbacks are that you tend to get long paths, which might keep you from testing some parts of the diagram until later in the project, when something goes wrong early in the test path.

Here is how to come up with a minimum path for the TFD in Figure 9.10. Start at the "IN" and take flow 1 to "NoGunNoAmmo." Then go to "HaveGun" via flow 2. Since flow 3 loops back to "HaveGun," take that next and then exit "HaveGun" via flow 4. The minimum path so far is 1, 2, 3, 4.

Now from "HaveGunHaveAmmo," go back to "HaveGun" via flow 5. Since flow 6 also goes from "HaveGunHaveAmmo" to "HaveGun," take flow 4 again and this time use flow 6 to return to "HaveGun." At this stage, the minimum path is 1, 2, 3, 4, 5, 4, 6, but there are still more flows to cover.

Take flow 7 out from "HaveGun" to go back to "NoGunNoAmmo." From here you can take flow 9 to "HaveGunHaveAmmo" and return back using flow 8. Now the path is 1, 2, 3, 4, 5, 4, 6, 7, 9, 8. All that remains now is to use the flows on the left side of the TFD.

You are at "NoGunNoAmmo" again so take flow 11 to "HaveAmmo" and then return to "NoGunNoAmmo" via flow 10. Only flow 12 and 13 are left now, so take 11 back to "HaveAmmo" where you can take 12 to "HaveGunHaveAmmo" and finally exit via flow 13 to the "OUT" box. The completed minimum path is 1, 2, 3, 4, 5, 4, 6, 7, 9, 8, 11, 10, 11, 12, 13. All thirteen flows on the TFD are covered in fifteen test steps.

There is usually more than one "correct" minimum path for any given TFD. For example, 1, 11, 10, 11, 12, 8, 9, 5, 7, 2, 3, 4, 6, 4, 13 is also a minimum path for the TFD in Figure 9.10. Diagrams that have more than one flow going to the OUT box will require more than one path. Even if you do not come up with the shortest path(s) mathematically possible, the purpose is to cover all of the flows in the least number of paths, which is one for the "Ammo" TFD.

Baseline Path Method

Baseline path generation begins by establishing as direct a path as possible from the "IN" terminator to the "OUT" terminator, which travels through as many states without repeating or looping back. This is designated as the *baseline path*. Additional paths are derived from the baseline by varying where possible, returning to the baseline path and following it to reach the "OUT" terminator. The process continues until all flows in the diagram are used at least once.

Baseline paths are more comprehensive than minimum paths, but still more economical than trying to cover every possible path through the diagram. They also introduce small changes from one path to another, so a game defect can be traced back to the operations that were different among the paths that passed and the one(s) that failed. One drawback of baseline paths is the extra effort to generate and execute the paths versus using the minimum path approach.

Still using the TFD in Figure 9.10, create a baseline path starting at the "IN" box and then traveling across the greatest number of states you can in order to get to the "OUT" box. Once you get to the "NoGunNo-Ammo" state from flow 1, the farthest distance to the "OUT" box is either through "HaveGun" and "HaveGunHaveAmmo" or through "HaveAmmo" and "HaveGunHaveAmmo." Take the "HaveGun" route by taking flow 2, followed by flow 4 and exiting through flow 13. This results in the baseline path of 1, 2, 4, 13.

The next thing to do is to branch wherever possible from the first flow on the baseline. These branches are called "derived" paths from flow 1. Flow 2 is already used in the baseline, so take flow 9 to "HaveGunHaveAmmo." From there flow 8 puts you back on the baseline path. Follow the rest of the baseline along flows 2, 4, and 13. The first derived path from flow 1 is 1, 9, 8, 2, 4, 13.

Continue to check for other possible branches after flow 1. Flow 11 comes out from "NoGunNoAmmo" and has not been used yet so follow it to "HaveAmmo." Then use flow 10 to return to the baseline. Finish this path by following the remainder of the baseline to the "OUT" box. This second path derived from flow 1 is 1, 11, 10, 2, 4, 13.

At this point there are no more new flows to cover from "NoGunNo-Ammo," so move along the next flow on the baseline which is flow 2. Stop here and look for unused flows to follow. You need to create a path using flow 3. Because it comes right back to the "HaveGun" state, continue along the remainder of the baseline to get to the path 1, 2, 3, 4, 13. The only other flow coming out of "HaveGun" is flow 7, which puts you right back on the baseline at flow 2. The final path derived from flow 2 is 1, 2, 7, 2, 4, 13.

Now on to flow 4! Flow 4 takes you to "HaveGunHaveAmmo," which has three flows coming out from it that are not on the baseline: 5, 6, and 8. We already used flow 8 in an earlier path, so there is no obligation to use it here. Flows 5 and 6 get incorporated into our baseline the same way because they both go back to the "HaveGun" state. The derived path using flow 5 is 1, 2, 4, 5, 4, 13 and the derived path from flow 6 is 1, 2, 4, 6, 4, 13.

It might seem as though you are done now because the next flow along the baseline goes to the "OUT" box, and you have derived paths from each other flow along the baseline. Upon further inspection, however, there is still a flow on the diagram that is not included in any of your paths: flow 12 coming from the "HaveAmmo" state. It is not connected to a state that is along the baseline, so it is easy to lose track of it. Pick up this flow by taking flows 1 and 11 to "HaveAmmo" and then use flow 12. You are now at "HaveGunHaveAmmo" and you must get back to the baseline to complete this path. Take flow 8, which is the shortest route and puts you back at "NoGunNoAmmo." Finish the path by following the rest of the baseline. This final path is 1, 11, 12, 8, 2, 4, 13.

As you can see, the baseline technique produces many more paths and results in much more testing time than a minimum path. The final baseline and derived paths for our "Ammo" TFD are as follows:

Baseline:

1, 2, 4, 13

Derived from flow 1:

1, 9, 8, 2, 4, 13

1, 11, 10, 2, 4, 13

Derived from flow 2:

1, 2, 3, 4, 13

1, 2, 7, 2, 4, 13

Derived from flow 4:

1, 2, 4, 5, 4, 13

1, 2, 4, 6, 4, 13

Derived from flow 11:

1, 11, 12, 8, 2, 4, 13

Expert Constructed Paths

Expert constructed paths are simply paths that a test or feature "expert" traces based on the expert's knowledge of how the feature is likely to fail or where she needs to establish confidence in a particular set of behaviors. They can be used by themselves or in combination with the minimum or baseline strategies. Expert constructed paths do not have to cover all of the flows in the diagram, nor do they have to be any minimum or maximum length. The only constraint is that, like all other paths, they start at "IN" and end at "OUT."

Expert paths can be effective at finding problems when there is organizational memory of what has failed in the past or what new game functions are the most sensitive. These paths could possibly have not shown up at all in a path list generated by the minimum or baseline criteria. The drawbacks of relying on this approach are the risks associated with not covering every flow and the possibility of tester bias producing paths that do not perform "unanticipated" sequences of events.

Some expert constructed path strategies are as follows:

- Repeat a certain flow or sequence of flows in combination with other path variations.
- Create paths that emphasize unusual or infrequent events.
- Create paths that emphasize critical or complex states.
- Create extremely long paths, repeating flows if necessary.
- Model paths after the most common ways the feature will be used.

For example, the "emphasize critical or complex states" strategy can be used for the "Ammo" TFD in Figure 9.10. In this case, the "HaveGun" state will be emphasized. This means that each path will pass through "HaveGun" at least once. It is also a goal to cover all of the flows with this path set. To keep the paths short, head for the "Exit" flow once the "HaveGun" state has been used.

One path that works is to go to "HaveGun," try to shoot, and then leave. This path would be 1, 2, 3, 4, 13. Another would incorporate the "DropGun" event in flow 7. The shortest way out from there is via flow 9 followed by 13, resulting in the path 1, 2, 7, 9, 13. You also need to include the two flows going into "HaveGun" from "HaveGunHaveAmmo." This produces the paths 1, 2, 4, 5, 4, 13 and 1, 2, 4, 6, 4, 13. Finish covering all of the flows leaving "HaveGunHaveAmmo" by using flow 8 in the path 1, 2, 4, 8, 9, 13.

All that remains are some slightly longer paths that cover the left side of the TFD. Flows 1, 11, 12 get you to "HaveGunHaveAmmo." The quickest way from there to "HaveGun" is either with flow 5 or 6. Choose flow 5, which results in the path 1, 11, 12, 5, 4, 13. You can eliminate or keep the earlier path that was made for the sole purpose of covering flow 5 (1, 2, 4,

5, 4, 13). It is no longer essential, since it has now also been covered by the path you needed for flow 12.

The last flow to cover is flow 10. Go to "HaveAmmo," take flow 10, go back through "HaveGun" and go out via flow 2. This gives you your final path of 1, 11, 10, 2, 4, 13. The list all of the paths that were just constructed for this set are as follows:

Expert path set:

1, 2, 3, 4, 13

1, 2, 7, 9, 13

1, 2, 4, 6, 4, 13

1, 2, 4, 8, 9, 13

1, 11, 12, 5, 4, 13

1, 11, 10, 2, 4, 13

Originally constructed but later eliminated:

1, 2, 4, 5, 4, 13

Combining Path Strategies

Testing uses time and resources that get more critical as the game project wears on. Here is one way to utilize multiple strategies that might make the best use of these resources for different stages of the project:

1. Use expert constructed paths early, even when the game is not yet code complete and everything might not be working. Limit yourself to paths that only include the parts that the developers are most interested in or paths that target the only parts of the game that are available for testing.
2. Use baseline paths to establish some confidence in the feature(s) being tested. This can begin once the subject of the TFD is feature complete. You might even want to begin by seeing if the game can pass the baseline path before trying to use the other paths in the set. Anything that fails during this testing can be narrowed down to a few test steps that vary between the failed path(s) and the successful ones.
3. Once the baseline paths all pass, use the minimum paths on an ongoing basis to ensure that it has not broken.

4. As any kind of delivery point nears, such as going to an investor demo, a trade show, or getting ready to go gold, revert back to the baseline or expert paths.

This puts a greater burden on the construction of the test paths, but over the course of a long project, it could be the most efficient use of the testers' and developers' time.

PRODUCING TEST CASES FROM PATHS

Here is how to produce a test case from a single TFD path. The subject of this example will again be the "Ammo" TFD in Figure 9.10. The test case will test getting ammo, then getting the gun, and then exiting. This is path 1, 11, 12, 13. To describe this test case, use the Data Dictionary definitions provided earlier in this chapter for the *Unreal Tournament* Bio-Rifle weapon.

Start constructing the test case with the Data Dictionary text for the "IN" box followed by the text for flow 1, which is the "Enter" flow:

Launch Unreal Tournament on the test PC.

Select a match and click the FIRE button to start the match.

Now add the text from the Data Dictionary for the "NoGunNoAmmo" state:

- ❑ *Check that the Bio-Rifle icon is empty in the graphical weapon inventory at the bottom of the screen.*
- ❑ *Check that the Bio-Rifle barrel is not rendered in front of your character.*
- ❑ *Check that you cannot select the Bio-Rifle weapon using the mouse wheel.*

Now take flow11 to get the Bio-Rifle ammo. Use the Data Dictionary entries for both the "GetAmmo" event and the "AmmoEffects" action:

Find a Bio-Rifle ammo pack on the floor in the arena and walk over it.

- ❑ *Check that the Bio-Rifle ammo sound is made.*

Flow 11 goes to the "HaveAmmo" state, so paste the "HaveAmmo" Data Dictionary text into the test case right after the text for flow 11:

- ❑ *Check that the Bio-Rifle icon is empty in the graphical weapon inventory at the bottom of the screen.*

- ❑ *Check that the Bio-Rifle barrel is not rendered in front of your character.*

- ❑ *Check that you cannot select the Bio-Rifle weapon using the mouse wheel.*

- ❑ *Check that the aiming reticle in the center of the screen has not changed.*

Next, add the text for the "GetGun" event and "GunEffects" action along flow 12:

Find an unloaded Bio-Rifle hovering above the floor of the arena and walk into it.

- ❑ *Check that the Bio-Rifle sound is made.*

- ❑ *Check that the game temporarily displays "You got the Bio-Rifle" in white text above the gun icons at the bottom of the screen.*

- ❑ *Check that the game simultaneously displays "Bio-Rifle" temporarily in blue text above the "You got the Bio-Rifle" message.*

- ❑ *Check that all temporary text on the display fades out slowly.*

Then paste the definition of the "HaveGunHaveAmmo" state:

- ❑ *Check that the Bio-Rifle icon is present in the graphical weapon inventory at the bottom of the screen.*

- ❑ *Check that the Bio-Rifle barrel is rendered in front of your character.*

- ❑ *Check that you can select the Bio-Rifle weapon using the mouse wheel.*

- ❑ *Check that the Bio-Rifle aiming reticle appears as a small blue broken triangle in the center of the screen.*

- ❑ *Check that the ammunition count in the right-hand corner of the screen is 40.*

Flow 13 is the last flow on the path. It is the "Exit" flow that goes to "OUT_GunAmmo." Complete the test case by adding the text for these two elements:

Hit the ESC key and exit the match.

At the main menu, click on "EXIT" to exit the game.

That is it! Here is how all of the steps look when they are put together:

Launch Unreal Tournament on the test PC.

Select a match and click the FIRE button to start the match.

- ❑ *Check that the Bio-Rifle icon is empty in the graphical weapon inventory at the bottom of the screen.*
- ❑ *Check that the Bio-Rifle barrel is not rendered in front of your character.*
- ❑ *Check that you cannot select the Bio-Rifle weapon using the mouse wheel.*

Find a Bio-Rifle ammo pack on the floor in the arena and walk over it.

- ❑ *Check that the Bio-Rifle ammo sound is made.*
- ❑ *Check that the Bio-Rifle icon is empty in the graphical weapon inventory at the bottom of the screen.*
- ❑ *Check that the Bio-Rifle barrel is not rendered in front of your character.*
- ❑ *Check that you cannot select the Bio-Rifle weapon using the mouse wheel.*
- ❑ *Check that the aiming reticle in the center of the screen has not changed.*

Find an unloaded Bio-Rifle hovering above the floor of the arena and walk into it.

- ❑ *Check that the Bio-Rifle sound is made.*

- ❑ *Check that the game temporarily displays "You got the Bio-Rifle" in white text above the gun icons at the bottom of the screen.*
- ❑ *Check that the game simultaneously displays "Bio-Rifle" temporarily in blue text above the "You got the Bio-Rifle" message.*
- ❑ *Check that all temporary text on the display fades out slowly.*
- ❑ *Check that the Bio-Rifle icon is present in the graphical weapon inventory at the bottom of the screen.*
- ❑ *Check that the Bio-Rifle barrel is rendered in front of your character.*
- ❑ *Check that you can select the Bio-Rifle weapon using the mouse wheel.*
- ❑ *Check that the Bio-Rifle aiming reticle appears as a small blue broken triangle in the center of the screen.*
- ❑ *Check that the ammunition count in the right-hand corner of the screen is 40.*

Hit the ESC key and exit the match.

At the main menu, click on "EXIT" to exit the game.

You can see how indenting the action and state definitions makes it easy to distinguish tester operations from things you want the tester to check for. When something goes wrong during this test, you will be able to document the steps that led up to the problem and determine what specifically was different from what you expected.

There are two techniques you can use to reuse this test case for another type of weapon. One is to copy the Bio-Rifle version and substitute the name of another weapon and its ammo type for "Bio-Rifle" and "Bio-Rifle ammo." This only works if all of the other details in the events, flows, and states are the same except for the gun and ammo names. In this case, Bio-Rifle-specific details were put into some of the definitions to give a precise description of what the tester should check.

"GunEffects" contains the following check, which references text color that varies by weapon. It is blue for the Bio-Rifle, but different for other weapons, such as red for the Rocket Launcher and white for the Minigun.

- ☐ *Check that the game simultaneously displays "Bio-Rifle" temporarily in blue text above the "You got the Bio-Rifle" message.*

Likewise, the "HaveGunHaveAmmo" state describes a specific color and shape for the Bio-Rifle aiming reticle as well as an ammunition count. Both vary by weapon type:

- ☐ *Check that the Bio-Rifle aiming reticle appears as a small blue broken triangle in the center of the screen.*

- ☐ *Check that the ammunition count in the right-hand corner of the screen is 40.*

This leaves you with the option to copy the Bio-Rifle Data Dictionary files into a separate directory for the new weapon. These files should then be edited to reflect the details for the new weapon type you want to test. Use those files to construct your test cases for the new weapon in the same way you did for the Bio-Rifle.

Remember that using text in the Data Dictionary is not your only option. You can also use screenshots or automated code. When executable code for each TFD element along a test path is pasted together you should end up with an executable test case. Use the "IN" definition to provide introductory code elements, such as including header files, declaring data types, and providing main routine opening braces. Use the "OUT" definition to perform cleanup actions such as freeing up memory, erasing temporary files, and providing closing braces.

Storing Data Dictionary information in separate files is not your only option. You could keep them in a spreadsheet or database and use a query to assemble the "records" for each TFD element into a report. The report could then be used for the manual execution of the game test.

TFD TEMPLATES

Appendix E provides ten TFD templates you can apply to various situations for a wide variety of games. You can recreate the diagrams with your own favorite drawing tool or use the SmartDraw (.sdf) templates among the companion files for this book. In the drawing files, suggested baseline paths are indicated by blue flows.

NOTE *All companion files for this book may be obtained by contacting the publisher at info@mercleaning.com.*

Flows in the template files are not numbered. There will be times when you will need to edit or otherwise customize the TFD to match the specific behaviors for your game. If you need an action and none is there, put in what you need. If there is an action on the TFD but you do not have one in your game, take the action out. Change the names of events, actions, or states to suit your game. Feel free to add any states you want to test that are not already provided. Once you have done all that, then add the flow numbers and define your paths.

WHEN SHOULD YOU CREATE A TFD?

Table 9.1 provides some guidelines for making a choice between using a combinatorial table or TFD for your test. If a feature or scenario has attributes that fall into both categories, consider doing separate designs of each type. For anything critical to the success of your game, create tests using both methods when possible.

TABLE 9.1 Test design methodology selection

Attribute/Dependency	Combinatorial	Test Flow Diagram
Game Settings	X	
Game Options	X	
Hardware Configuration	X	
Game State Transitions		X
Repeatable Functions		X
Concurrent States	X	
Operational Flow		X
Parallel Choices	X	X
Story Paths or Game Routes		X

Test Flow Diagrams are used to create models of how the game should work from the player's perspective. By exploring this model, the tester can create unanticipated connections and discover unexpected game states. TFDs also incorporate invalid and repetitive inputs to test the game's behavior. TFD tests will demonstrate if expected behavior occurs and unexpected behavior does not. Complex features can be represented by complex TFDs, but a series of smaller TFDs is preferred. Good TFDs are the result of insight, experience, and creativity.

EXERCISES

1. Describe how you would apply the "Ammo" TFD in Figure 9.10 to an archer in an online role-playing game. Include any modifications you would make to the TFD structure as well as to individual states, events, or actions.
2. Update the diagram in Figure 9.10 to account for what happens when the player picks up ammo that does not match the type of gun he has.
3. Create a set of baseline and minimum paths for the updated TFD you created in Exercise 2. Create Data Dictionary entries and write out the test case for your minimum path. Reuse the Data Dictionary entries already provided in this chapter and create any new Data Dictionary entries you need.
4. Construct a TFD for a mobile game that is suspended when the user receives a call or the screen is locked due to inactivity. Try to keep the number of states low. The game should be resumed once the call ends or the user unlocks the screen.

CHAPTER 10

CLEANROOM TESTING

Cleanroom testing is a technique extracted from a software development practice known as *Cleanroom Software Engineering*. The original purpose of Cleanroom testing was to exercise software to make mean time to failure (MTTF) measurements over the course of a project (Prowell et al. 1999).

In this chapter, Cleanroom testing is applied to the problem of why players find bugs in games that have been through thousands of hours of testing before being released. If one measure of a game's success is that the players will not find any bugs, then the game team's test strategy should include a way to detect and remove the defects that are most likely to be found by users.

In earlier chapters, we discussed how oftentimes QA will find defects that the game's publisher will decide to "let go," and release the game knowing that those bugs are still in the code.

Sometimes the best QA teams and the most well-intended publishers will release a game with a bug (or bugs) that only get found by players. So how do players find defects that testers missed? Users find defects in games by playing it the way players play it. That is a little bit of a tongue twister, but it points to a testing approach that exercises the game according to the way the players are going to play it. That is what Cleanroom test development does; it produces tests that play the game the way players will play it.

USAGE PROBABILITIES

Usage probabilities, also referred to as *usage frequencies*, tell testers how often game functions should be used in order to realistically mimic the way players will use the game. They can be based on actual data you might have from studies of game players or based on your own expectations about how the game will be played. They should also take into account the possible evolution of a player's play style during the life of the game. A player's patterns would be different just after running the tutorial than they would be by the time the player reaches the boss on the final level. Initially, the player would utilize fundamental operations and have few, if any, special items unlocked. Clicking menu icons would occur more frequently than key commands and user-defined macros. Matches or races might take longer at the end of the game due to the higher difficulty and closer matching of the player's skill to that of the in-game opponents. Usage information can be defined and utilized in three different ways:

- mode-based usage
- player-type usage
- real-life usage

Mode-Based Usage

Game usage can change based on which mode the player is using, such as single-player, campaign, multiplayer, or online.

Single-player mode could involve one or only a few confrontations or missions. The action usually starts right away so the player is less likely to perform "build-up" operations such as building advanced units and spending money or points on expensive skill-boosting items. Some features might not be available at all to the single-player, such as certain characters, weapons, or vehicles. The single-player's character might also have limited race, clan, and mission options.

Campaigns tend to start the player with basic equipment and opponents and then introduce more and more complex elements as the campaign progresses. For sports games, Franchise or Season modes provide unique options and experiences that are not available when playing a single game, such as draft picks, training camp, trading players, and negotiating salaries. RPG games will provide more powerful spells, armor, weapons, and opponents as your characters level up. Racing games can provide more powerful vehicles, add-ons, and power-ups, as well as more challenging tracks.

Multiplayer gaming can take place on the same machine, across two connected consoles, or over the Internet for more massive multiplayer experiences. Headset accessories are used for team confrontations but are not something you are likely to use by yourself unless the game has voice commands. Text chatting is used in multiplayer PC games, giving the text keyboard a workout. Game controls can also be assigned to taunts to adversaries or "pings" to teammates, as in *Apex Legends*. Because online multiplayer games can have the ability to bring players together from around the globe, game sessions might involve a variety of time zone-based game clocks, and language settings.

Player-Type Usage

Another factor that influences game usage is the classification of four multi-user player categories described by Richard A. Bartle in "Hearts, Clubs, Diamonds, Spades: Players Who Suit MUDs" (Bartle 1996). He describes players by their tendencies to emphasize either Achievement, Exploration, Socializing, or Killing when they participate in multiplayer games.

Achievers want to complete game goals, missions, and quests. They will gain satisfaction in advancing their characters' level, point, and money totals in the most efficient way possible. Achievers often replay the game at a higher level of difficulty or under difficult circumstances such as using a last-place team or going into combat armed only with a knife. They will also be interested in reaching bonus goals and completing bonus missions.

Explorers are interested in finding out what the game has to offer. They will travel around to find obscure places and the edges of the map; unmapped territory will draw their attention. Explorers will look around and appreciate the art and special beauty in the game such as a particularly nice moonrise or light shining through a cathedral window. They are also likely to attempt interesting features, animations, combinations (*combos*), and physics effects. Expect the Explorer to try to open every door and check the inventory at all of the stores. The Explorer wants to figure out how things work. (For this type of player, think of the phrase "I wonder what would happen if…?")

The goal of the *Socializer* is to use the game as a means to role play and get to know other players. Chat and messaging facilities are important to them, as well as joining social groups within the game such as clans, guilds, and so on. They might host meetings or tournaments, or bring many players together in one place for announcements, trading, or even an occasional wedding. Socializers will use special game features once they find out about them from other players.

Killers enjoy getting the best of other players. They engage in player versus player and real versus realm battles. Killers know where the taunt keys are and how to customize and activate an end zone celebration. Headsets, chats, and private messages are also tools that the Killer uses to bait and humiliate his opponents.

Finally, here are some other player types to consider when you go to test your game the same way players play the game:

- ***Casual player:*** Sticks mostly to functions described in the tutorial, user manual, and on-screen user interface.
- ***Hardcore player:*** Uses function keys, macros, turbo buttons, and special input devices such as joysticks and steering wheels. Checks the Internet for tricks and tips. Might also have advanced hardware capable of running games at the highest possible graphics resolution and frame rate. MMORPGs and RTS games such as *World of Warcraft* and *Age of Empires* are frequent hunting grounds for this gamer type, some of whom might operate multiple characters and computers at the same time.
- ***Button Masher:*** Values speed and repetition over caution and defense. Quickly degrades the A button on the controller to run faster, jump higher, or strike first. Will run out of ammo. Touch-enabled smartphones are used by another type of player much like the Button Masher: the ***Screen Scratcher***.
- ***Customizer:*** Uses all of the game's customization features and plays the game with custom elements. Will also incorporate unlocked items, decals, skins, tags, teams, and variant cards.
- ***Exploiter:*** Always looking for a shortcut. Will use cheat codes, look for cracks in zone walls, and pick off opponents from a secret or unreachable spot. Creates bots to craft items, earn points, and level up. Uses infinite card combos against AI and human Collectible Card Game (CCG) opponents.

Real-Life Usage

Some games have built-in mechanisms for capturing your in-game actions. This data might be available to you within the game, published on your game profile for other players to see, or even sent as raw data back to the game publisher for analysis or debugging. Capturing data from real-world gaming "friends" or in-game NPCs, such as coaches, goblins, and rival drivers, can be used to let you practice against their style of play so you can defeat them the next time you battle for real. The 2015 edition of *Madden NFL* puts such usage information into the player's hands by incorporating real-time data into each play.

One example is an on-screen "Previous Play" information box that compares the player's yards gained versus the results from all *Madden* online players at that moment. A player may lose four yards on their play, but will see that the community of Madden players who ran that play averaged a gain of 10.1 yards.

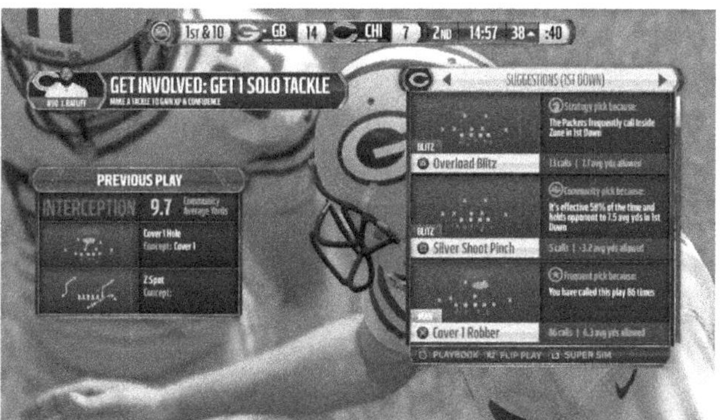

FIGURE 10.1 *Madden NFL 15* Previous Play Information Box

Armed with this kind of information—whether aggregated by the publisher or from individual players' results—testers can exercise games according to different player styles and the expected frequency of use of various game actions, choices, and options. It is important to account for these tendencies because testing based entirely on balanced use of the game features might not reveal such defects a memory overflow caused by tapping the A button repeatedly on every play over the course of a maximum length game.

CLEANROOM TEST GENERATION

It is possible to generate Cleanroom tests using any of the methods covered in this book. You can also spontaneously create your own Cleanroom tests. A usage probability must be assigned to each step in the test. This can be done in writing or you can keep track in your head. Use the usage probability to select test steps, values, or branches, and put them in sequence to produce tests that reflect your usages. For example, if you expect a simulation game player to develop residential property 50% of the time, commercial property 30% of the time, and industrial property 20% of the time, then your Cleanroom tests should reflect those same frequencies.

Cleanroom Combinatorial Tables

Cleanroom combinatorial tables will not necessarily be "pairwise" combinatorial tables (see Chapter 8, "Combinatorial Testing"). The number of tests to be created is determined by the test designer, and the values for each test will be chosen on the basis of their frequency of use rather than whether they satisfy one or more necessary value pairs.

To produce Cleanroom combinatorial tables, assign usage probabilities to the test values of each parameter. The probabilities of the set of values associated with a single parameter must add up to 100%.

To illustrate how this is done, revisit the parameter and value choices for the *Halo: Reach* Advanced Controls table that you completed in Chapter 8, Table 8.24. The test values for each parameter are listed below with the default values identified:

- *Look Sensitivity*: 1, 3 (default), 10
- *Look Inversion*: Inverted, Not Inverted (default)
- *AutoLook Centering*: Enabled, Disabled (default)
- *Crouch Behavior*: Hold to Crouch (default), Toggle
- *Clench Protection*: Enabled, Disabled (default)

Next, usage percentages need to be determined for each of the table's parameters. If you are considering testing against more than one player profile, you can make a separate usage table for each parameter with a column of usage percentages for each of the profiles you intend to test. Tables 10.1 through 10.5 show multiple profile usages for each of the five *Halo: Reach* Advanced Controls parameters that you will incorporate into your Cleanroom combinatorial table.

NOTE

This chapter presents a variety of usage numbers to illustrate differences among user types based on personal experience. If you have data gathered through scientific means, then that is what you should use instead. If these numbers do not make sense to you, then please consider them "for educational purposes only" as you continue through the examples in this chapter.

Beginning with Table 10.1, distinct usage percentages for the Look Sensitivity parameter are provided to reflect the expected tendencies for each of the depicted player types.

TABLE 10.1 Look Sensitivity values with usage percentages

Look Sensitivity	Casual	Achiever	Explorer	Multiplayer
1	10	0	10	5
3	85	75	70	75
10	5	25	20	20
TOTAL	100	100	100	100

Table 10.2 provides a different set of usage values for the "Inverted" and "Not Inverted" options available for the Look Inversion parameter.

TABLE 10.2 Look Inversion values with usage percentages

Look Inversion	Casual	Achiever	Explorer	Multiplayer
Inverted	10	40	30	50
Not Inverted	90	60	70	50
TOTAL	100	100	100	100

Table 10.3 introduces a situation where the "Disabled" value for Auto-Look Centering has a 100% weighting. As a consequence, it will be tested using this same value throughout the entire set of Cleanroom tests generated for the Achiever player type.

TABLE 10.3 AutoLook Centering values with usage percentages

AutoLook Centering	Casual	Achiever	Explorer	Multiplayer
Enabled	30	0	20	10
Disabled	70	100	80	90
TOTAL	100	100	100	100

Table 10.4 provides Crouch Behavior usage values that are mostly biased toward the "Hold" value, except for the Explorer, who is given an equal probability of selecting either the "Hold" or "Toggle" option.

TABLE 10.4 Crouch Behavior centering values with usage percentages

Crouch Behavior	Casual	Achiever	Explorer	Multiplayer
Hold	80	75	60	90
Toggle	20	25	50	10
TOTAL	100	100	100	100

Table 10.5 contains the final set of probabilities that are needed to proceed with Cleanroom test generation.

TABLE 10.5 Clench Protection values with usage percentages

Clench Protection	Casual	Achiever	Explorer	Multiplayer
Enabled	25	60	50	90
Disabled	75	40	50	10
TOTAL	100	100	100	100

Use Tables 10.1 through 10.5 as you work through the tutorial below to complete your first Cleanroom combinatorial table.

Cleanroom Combinatorial Examples

A Cleanroom combinatorial table can be constructed for any of the player usage profiles you define. For this example, you will create one such table for the Casual player. To decide which value to choose for each parameter, you need a random number source. You could think of a number in your head, write a program to generate a list of numbers, or roll electronic dice on your smartphone. Many spreadsheet programs have functions that can generate random numbers. Perhaps the quickest way is to visit *https://www.random.org*, and use the True Random Number Generator tool at the top of the page. There is no wrong way as long as the number range is from 1–100, and selection is not biased toward any portion of the range.

Start building the table with an empty template that has column headings for each of the parameters. Decide how many tests you want, and leave room for them in the table. A Cleanroom combinatorial table "shell" for the *Halo: Reach* Advanced Controls is shown in Table 10.6. It has room for six tests.

TABLE 10.6 *Halo: Reach* Advanced Controls Cleanroom Combinatorial Table Shell

Test	Look Sensitivity	Look Inversion	AutoLook Centering	Crouch Behavior	Clench Protection
1					
2					
3					
4					
5					
6					

Because there are five parameters, get five random numbers in the range of 1–100. These will be used one at a time to determine the values for each parameter in the first test. Construct the first test from the five numbers 30, 89, 13, 77, and 25.

Referring back to Table 10.1, the Casual player is expected to set the Look Sensitivity to "1" 10% of the time, to "3" 85% of the time, and to "10" 5% of the time. Assigning successive number ranges to each choice results in a mapping of 1–10 for Look Sensitivity = 1, 11–95 for Look Sensitivity = 3, and 96–100 for Look Sensitivity = 10. The first random number, 30, falls into the 11–95 range, so enter "3" in the first column of the test table.

Likewise, Table 10.2 provides a range of 1–10 for Look Inversion = Inverted and 11–100 for Look Inversion = Not Inverted. The second random number is "89," which is within the 11–100 range. Enter "Not Inverted" in the Look Inversion column for Test 1.

In Table 10.3, the AutoLook Centering usage ranges for the Casual player are 1–30 for "Enabled" and 71–100 for "Disabled." The third random number is "13," so enter "Enabled" in Test 1's AutoLook Centering column.

Table 10.4 defines an 80% usage for Crouch Behavior = Hold and a 20% usage for "Toggle." The fourth random number is "77," which is within the 1–25 range for the "Yes" setting. Enter "Hold" in the Crouch Behavior column for Test 1.

Last, Table 10.5 defines the Clench Protection Casual player usage as 25% for "Enabled" and 75% for "Disabled." The last random number is 25, which is within the 1–25 range for the Enabled setting. Complete the definition of Test 1 by putting "Enabled" in the Clench Protection column for Test 1.

Table 10.7 shows the first test constructed from the random numbers 30, 89, 13, 77, and 25.

TABLE 10.7 The First Advanced Controls Cleanroom combinatorial test

Test	Look Sensitivity	Look Inversion	AutoLook Centering	Crouch Behavior	Clench Protection
1	3	Not Inverted	Enabled	Hold	Enabled

A new set of five random numbers is required to produce the second test case. Use 79, 82, 57, 27, and 8.

The first number is "79," which is within the 11–95 range for Look Sensitivity = 3. Put a "3" again in the first column for Test 2. The second usage number is "82." It falls within the 11–100 range for Look Inversion = Not Inverted, so put "Not Inverted" that column for Test 2. Your third random number is "57." This number is in the 31–100 range for AutoLook Centering, so enter "Disabled" into that column for Test 2. The fourth usage number is "27." This is within the 1–80 range for Crouch Behavior = Hold. Add "Hold" to the fourth column of values for Test 2. The last random number is "8." This usage value corresponds to the "Enabled" value range of 1–25 for the Clench Protection parameter. Complete Test 2 by entering "Enabled" in the last column. Table 10.8 shows the first two completed rows for this Cleanroom combinatorial table.

TABLE 10.8 Two Advanced Controls Cleanroom combinatorial tests

Test	Look Sensitivity	Look Inversion	AutoLook Centering	Crouch Behavior	Clench Protection
1	3	Not Inverted	Enabled	Hold	Enabled
2	3	Not Inverted	Disabled	Hold	Enabled

The third test in this table is constructed from the random number sequence 32, 6, 11, 64, and 66. Once again, the first value corresponds to the default Look Sensitivity value of "3." The second usage number is "6," which results in the first "Inverted" entry for the Look Inversion parameter by virtue of being inside the 1–10 range for that value. The third random number for Test 3 is "11," which gives you an "Enabled" value for the Auto-Look Centering parameter. The number to use for determining the Crouch Behavior test value is "64," which maps to the 1–80 range for the "Hold" choice. The fifth number provides another "first"—a "Disabled" value for Clench Protection, because it falls within the 26–100 range. Table 10.9 shows the first three tests entered in the table.

TABLE 10.9 Three Advanced Controls Cleanroom combinatorial tests

Test	Look Sensitivity	Look Inversion	AutoLook Centering	Crouch Behavior	Clench Protection
1	3	Not Inverted	Enabled	Hold	Enabled
2	3	Not Inverted	Disabled	Hold	Enabled
3	3	Inverted	Enabled	Hold	Disabled

Continue by using the random numbers 86, 64, 22, 95, and 50 for Test 4. The "86" is within the 11–95 range for Look Sensitivity =3, so put a "3" again in column one. A "64" is next in the usage number list. It maps to the "Not Inverted" range for Look Inversion. The next number, "22," corresponds to "Enabled" for AutoLook Centering. The "95" provides the first Toggle value for Crouch Behavior in this set of tests. The Clench Protection number is "50," which puts another "Disabled" value in that column. Table 10.10 shows the table with four of the six tests defined.

TABLE 10.10 Four Advanced Controls Cleanroom combinatorial tests

Test	Look Sensitivity	Look Inversion	AutoLook Centering	Crouch Behavior	Clench Protection
1	3	Not Inverted	Enabled	Hold	Enabled
2	3	Not Inverted	Disabled	Hold	Enabled
3	3	Inverted	Enabled	Hold	Disabled
4	3	Not Inverted	Enabled	Toggle	Disabled

Your fifth set of random numbers is 33, 21, 76, 63, and 85. The 33 puts a "3" in the Look Sensitivity column. The "21" is in the "Not Inverted" range for Look Inversion. An "85" is within the "Disabled" range for AutoLook Centering. The "63" corresponds to a "Hold" value for Crouch Behavior and the "85" causes another "Disabled" to be put in the last column for the Clench Protection parameter. Table 10.11 shows the Cleanroom combinatorial table with five tests defined. Only one more to go now!

TABLE 10.11 Five Advanced Controls Cleanroom combinatorial tests

Test	Look Sensitivity	Look Inversion	AutoLook Centering	Crouch Behavior	Clench Protection
1	3	Not Inverted	Enabled	Hold	Enabled
2	3	Not Inverted	Disabled	Hold	Enabled
3	3	Inverted	Enabled	Hold	Disabled
4	3	Not Inverted	Enabled	Toggle	Disabled
5	3	Not Inverted	Disabled	Hold	Disabled

One more number set is needed to complete the table. Use 96, 36, 18, 48, and 12. The first usage number of "96" is high enough to be in the 96-100 range for the "10" Look Sensitivity value. This marks the first time that value appears in the table. Moving through the rest of the numbers, the "36" puts a "No" in the Look Inversion column, "18" corresponds to AutoLook Centering = Enabled, "48" is in the range for Crouch Behavior = Hold, and "12" completes the final test row with "Enabled" for Clench Protection. Table 10.12 shows all six Cleanroom combinatorial test cases.

TABLE 10.12 Completed Advanced Controls Cleanroom combinatorial tests

Test	Look Sensitivity	Look Inversion	AutoLook Centering	Crouch Behavior	Clench Protection
1	3	Not Inverted	Enabled	Hold	Enabled
2	3	Not Inverted	Disabled	Hold	Enabled
3	3	Inverted	Enabled	Hold	Disabled
4	3	Not Inverted	Enabled	Toggle	Disabled
5	3	Not Inverted	Disabled	Hold	Disabled
6	10	Not Inverted	Enabled	Hold	Disabled

Your keen testing eye should have noticed that Look Sensitivity = 1 was never generated for this set of tests. That is a function of its relatively low probability (10%), the low number of test cases that you produced, and the particular random number set that was the basis for selecting the values for table. In fact, if you stopped generating tests after five test cases instead of six, the default value of "3" would have been the only value for Look Sensitivity that appeared in the table. This should not be considered a problem for a table of this size. If a value has a 5% or higher usage probability and you do not see it at all in a test set of 100 or more tests, then you can suspect that something is wrong with either your value selection process or your random number generation.

Also notice that some values appear more frequently or less frequently than their usage probability would suggest. AutoLook Centering = Enabled has only a 30% usage for the Casual profile, but it appears in 67% (4/6) of the tests generated. This is mainly due to the low number of tests created for this table. With a test set of 50 or more, you should see a better match between a value's usage probability and its frequency in the test set.

Just to reinforce the fact that the Cleanroom combinatorial table method does not guarantee it will provide all test value pairs that are required for a pairwise combinatorial table, confirm that the pair AutoLook Centering = Disabled and Crouch Behavior = Toggle is absent from Table 10.12. Now take a moment to see which other missing pairs you can find.

You will recall that pairwise combinatorial tables are constructed vertically, one column at a time. Until you complete the process for building the table you do not know what the test cases will be nor how many tests will result. Because Cleanroom combinatorial tables are constructed horizontally—one line at a time—you get a completely defined test on the very first row, and every row after that for as many Cleanroom combinatorial tests as you choose to produce.

TFD Cleanroom Paths

Cleanroom TFD tests come from the same diagram you use for creating minimum, baseline, and expert constructed paths. Cleanroom test paths travel from state to state by choosing each subsequent flow based on its usage probability.

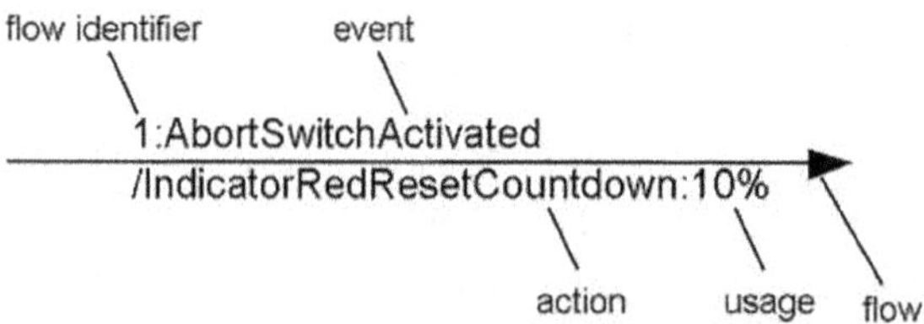

FIGURE 10.2 Example flow with usage probability

A usage probability must be added to each flow if the TFD is going to be used for Cleanroom testing. The probabilities of the set of flows exiting each state must add up to 100%. Figure 10.2 shows a flow with the usage probability after the action. If there is no action on the flow, then the usage probability gets added after the event.

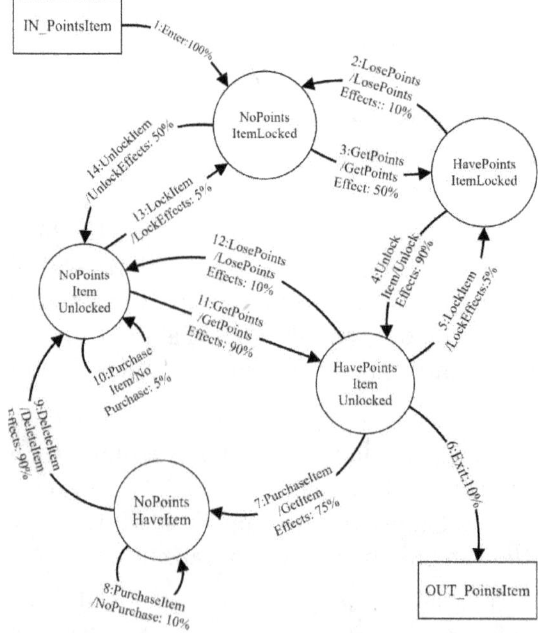

FIGURE 10.3 Unlock Item TFD with usage probabilities added

Figure 10.3 shows an entire TFD with flow numbers and usage percentage amounts. Remember, the probabilities of flows exiting each state must add up to 100%. You might recognize this TFD from the templates provided in Appendix D. The flow numbers and usage percentages make this TFD ready for Cleanroom testing.

TFD Cleanroom Path Example

With the usage information added to the TFD, generate random numbers to guide you around the diagram from flow to flow until you reach the "OUT" terminator. The resulting path defines a single test. Continue generating as many paths as you like, using new random number sets each time. Experience has shown that it is a good practice to always assign a 10% value to the "Exit" flow. A larger value will result in paths that exit too soon and a smaller value will cause too many paths that seem to go on forever before finally exiting. The 10% value provides a nice mix of long, medium, and short paths in your Cleanroom test set.

Each Cleanroom test case is described by the sequence of flow numbers along the Cleanroom path. Because the path length can vary from one test to another, you will not know ahead of time how many random numbers you need to generate for all of your paths. The result is that you could exit some tests after only a few flows, or you could travel around the diagram several times before reaching the "OUT" box. Normally, you would generate the random numbers as you need them, but for your convenience the random number set for the example in this section is 30, 27, 35, 36, 82, 59, 92, 88, 80, 74, 42, and 13.

Generating a test case for the TFD in Figure 10.4 starts at the "IN" box. The only flow from there has a 100% usage, so there is no need to produce a random number. You *must* begin your test with this flow. Next, there are two possible ways out from the "NoPointsItemLocked" state: flow 3 and flow 14. Each of those flows has the usage probability of 50%. Assign them each a random number range according to their numerical order. Use flow 3 if the random number is 1–50 and use flow 14 if it is 51–100. Get the random number "30" from the list above, and take flow 3 to "HavePointsItem-Locked." The test path so far is 1, 3.

There are two flows exiting the state "HavePointsItemLocked." Flow 2 has a 10% usage, and flow 4 has a 90% usage. The range for flow 2 is 1–10 and for flow 4 it is 11–100. Use "27" as the random number for this flow. That sends the test along flow 4 to "HavePointsItemUnlocked." The test path at this point is 1, 3, 4.

"HavePointsItemUnlocked" is the most interesting state so far, with four flows to choose from for the next step in your test. Flow 5 has a 5% usage, flow 6 has 10%, flow 7 has 75%, and flow 12 has 10%. The corresponding number ranges are 1–5 for flow 5, 6–15 for flow 6, 16–90 for flow

7, and 91–100 for flow 12. The next random number is "35." Your test path now takes flow 7 to "NoPointsHaveItem." The path is now 1, 3, 4, 7.

From "NoPointsHaveItem" there are two flow choices: flow 8 with a 10% usage and flow 9 with a 90% usage. You will take flow 8 if the random number is in the range 1–10 and flow 9 if it is within 11–100. Your new random number is "36," so take flow 9 to "NoPointsItemUnlocked." The test path is currently 1, 3, 4, 7, 9.

Flows 10, 11, and 13 all leave "NoPointsItemUnlocked." Flow 10's usage is 5% (1–5), flow 11 has a 90% usage (6–95), and flow 13 has a 5% (96–100) usage. Another random number is generated and it is "82." That is within the range for flow 11, so take that flow to "HavePointsItemUnlocked." The path has grown to 1, 3, 4, 7, 9, 11, but you are not done yet.

You are back at "HavePointsItemUnlocked," and the next random number is 59. That fits in the 16–90 range for flow 7, taking you on another trip to "NoPointsHaveItem." A usage of "92" here matches up with flow 9, going to "NoPointsItemUnlocked." The test path is now 1, 3, 4, 7, 9, 11, 7, 9.

The next random number is "88." This takes you from "NoPointsItemUnlocked" to "HavePointsItemUnlocked" via flow 11. The "80" takes you along flow 7 for the third time in this path, and the next number, "74," sends you to "NoPointsItemUnlocked" via flow 9. A "42" in the random number list chooses flow 11, which brings you once again to "HavePointsItemUnlocked." These flows extend the path to 1, 3, 4, 7, 9, 11, 7, 9, 11, 7, 9, 11.

The next random number to use is "13." This falls within the 6–15 range, which corresponds to flow 6. That is the "Exit" flow, which goes to the "OUT" terminator. This marks the end of this test path. The completed path is 1, 3, 4, 7, 9, 11, 7, 9, 11, 7, 9, 11, 6.

Once a path is defined, create the test cases using the Data Dictionary techniques described in Chapter 11. To create an overview of this test, list the flows, actions, and states in the order they appear along the path. List the flow number for each step in parentheses at the beginning of each line, as follows:

IN_PointsItem

(1) Enter, NoPointsItemLocked

(3) GetPoints, GetPointsEffects, HavePointsItemLocked

(4) UnlockItem, UnlockEffects, HavePointsItemUnlocked

(7) PurchaseItem, GetItemEffects, NoPointsHaveItem

(9) DeleteItem, DeleteItemEffects, NoPointsItemUnlocked

(11) GetPoints, GetPointsEffects, HavePointsItemUnlocked

(7) PurchaseItem, GetItemEffects, NoPointsHaveItem

(9) DeleteItem, DeleteItemEffects, NoPointsItemUnlocked

(11) GetPoints, GetPointsEffects, HavePointsItemUnlocked

(7) PurchaseItem, GetItemEffects, NoPointsHaveItem

(9) DeleteItem, DeleteItemEffects, NoPointsItemUnlocked

(11) GetPoints, GetPointsEffects, HavePointsItemUnlocked

(6) Exit, OUT_PointsItem

Generating this path provided some expected results. The path starts with the "IN" and ends with the "OUT," which is mandatory. Flows with large percentages were selected often, such as flows 9 and 11, which each have a 90% usage probability.

Did anything surprise you? Some flows and states did not appear in this path at all. That is acceptable for a single path. When you create a set of paths, you should expect to explore a wider variety of flows and states.

Was the flow longer than you expected? Flows 7, 9, and 11 appeared multiple times in this path. This is not what you would expect from minimum or baseline path sets. It is also interesting to note that those three flows form a loop. They were used three times in a row before finally exiting and ending the path.

Was the path longer than you wanted it to be? Is this a path you would have chosen on your own? Because this technique is based on a process rather than the ideas or preconceptions of a particular tester, the paths are free of bias or limitations. Cleanroom paths also highlight the fact that the game is not played one operation at a time and then turned off. These paths will test realistic game-use scenarios if your percentages are reasonably correct. As a result, your Cleanroom tests will have the ability to reveal defects that are likely to occur during extended or repeated game use.

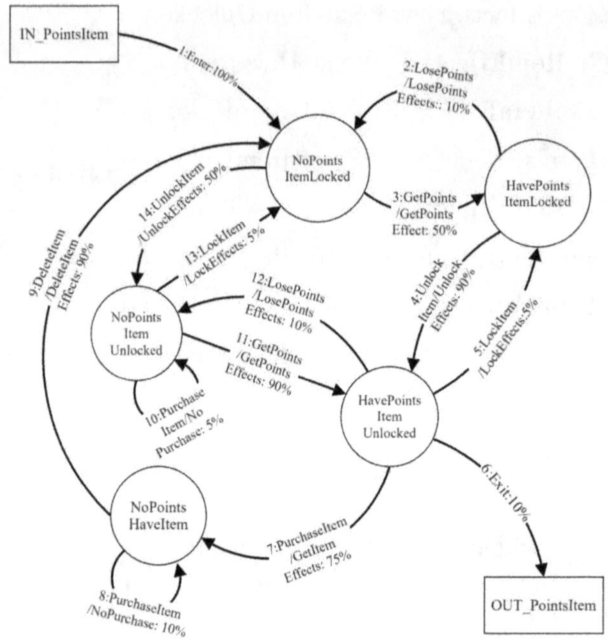

FIGURE 10.4 Unlock Item TFD with altered flow 9

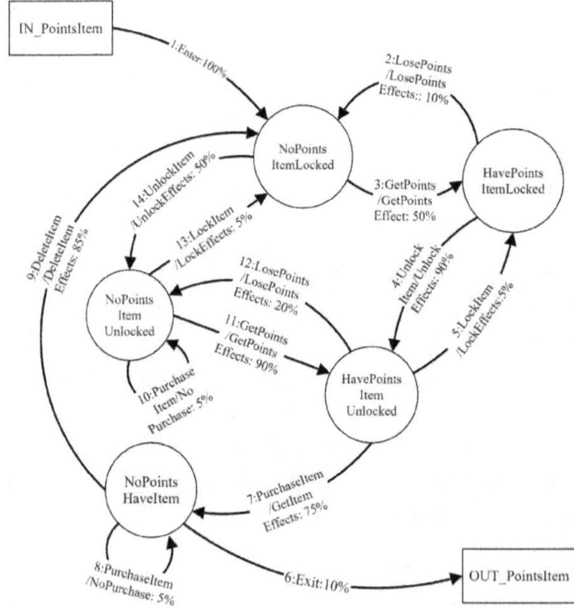

FIGURE 10.5 Unlock Item TFD with altered flows 6 and 9

Flow Usage Maintenance

Eventually, you will need to move one or more flows around on your TFD. This could perhaps affect your usage values. When a flow's destination (arrowhead end) changes, you are not required to change its usage. Conversely, if you change a flow to originate from a new state, you must re-evaluate the usage values for all flows coming from both the new state and the original one.

Figure 10.4 shows an updated version of the Unlock Item TFD. Flow 9 on the left side of the diagram now goes all the way back to "NoPointsItemLocked" instead of "NoPointsItemUnlocked." The usage percentage for flow 9 does not have to change. The percentages for all the flows coming from "NoPointsHaveItem" still add up to 100: 10% for flow 8 and 90% for flow 9.

Figure 10.5 includes a second update to the "Unlock Item" TFD.

Flow 6 originally started at "HavePointsItemUnlocked," but now it goes from "NoPointsHaveItem" to the "OUT" box. For this case, all flows coming from both "HavePointsItemUnlocked" and "NoPointsHaveItem" were re-evaluated to add up to 100% from each originating state.

For "HavePointsItemUnlocked," one or more percentages need to increase because that state lost a flow. You can give flow 12 the 10% that used to be allocated to flow 6. That would not overly inflate the usage for flow 7, and it keeps flow 5's usage small. As Figure 10.6 shows, flow 12 now has a 20% usage instead of its original 10% value.

Additionally, one or more flows coming from "NoPointsHaveItem" must now be reduced to make room for the new flow. Because flow 6 is an "Exit" flow, it must have a 10% usage. Two other flows come from "NoPointsHaveItem:" flow 8 with a 10% usage and flow 9 with a 90% usage. Reducing flow 8 by 10% will put it at 0%, meaning it will never be selected for any Cleanroom paths for this TFD. Instead, take away 5% from flow 8 and 5% from flow 9. The new percentages for these flows are reflected in Figure 10.6. Alternatively, you could have taken 10% away from flow 9 and left flow 8 at 10%. Your choice depends on what distribution you think best reflects the expected relative usage of these flows according to the game player, mode, or data you are trying to model.

Flow Usage Profiles

You might want to have multiple usage profiles to choose from when you create TFD Cleanroom paths. One way to accomplish this is to create copies of the TFD and change the usage numbers to match each profile. Another solution is to do what you did for combinatorial profiles: produce a mapping between each test element and its usage probability for one or more game users, types, or modes. In this case, usage numbers should not appear on the TFD. Figure 10.7 shows the Unlock Item TFD without usage percentages on the flows.

Table 10.13 shows how one profile's probabilities map to the flows on the TFD. Document the random number range that corresponds to each flow's usage. For example, because flows 3 and 14 go out from "NoPointsItemLocked," flow 3 gets the range 1–50 and flow 14 gets 51–100. When you edit the TFD to add, remove, or move flows, you must revisit this table and update the usage and range data.

TABLE 10.13 Casual Player usage table for Unlocked Item TFD flows

Flow	Casual
1	100
2	10
3	50
4	90
5	5
6	10
7	75
8	10
9	90
10	5
11	90
12	10
13	5
14	50
TOTAL	600

The total at the bottom of the flow probability table is a good way to check that your percentages add up correctly. The total should be equal to 100 (for the "Enter" flow) plus 100 times the number of states on the diagram (flows exiting each state must add up to 100%). The TFD in Figure 10.6 has five states, so "600" is the correct total.

Generate your Cleanroom tests from the flow usage table similarly to the way you do when the flow usage is on the diagram. The only difference is the extra step to look up the flow's range in the table. If you are creating an automated process or tool to construct TFD Cleanroom paths, this table could be stored in a database or exported to a text file.

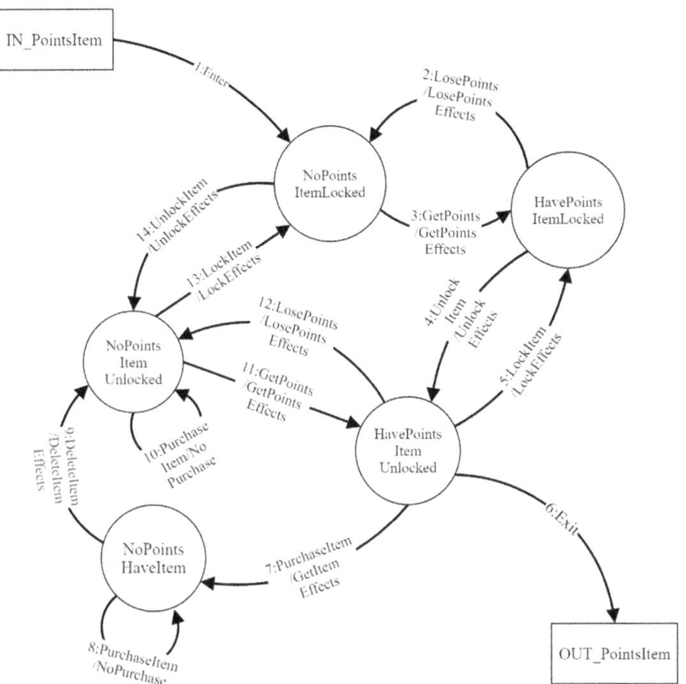

FIGURE 10.6 Unlock Item TFD without usage probabilities

Careful Testing Requires Careful Testing!
Admittedly, keeping track of flow usage in a table presents some problems. Because the flow numbering does not have to be related to the way flows appear on the diagram, it takes a little more work to identify the flows coming from each individual state. For example, the flows coming from "NoPointsItemLocked"—3 and 14—are at opposite ends of the flow list. This issue can become more of a problem when many flows are added, moved, or removed to adapt to changes in the game software. Just be careful, and check your numbers when you are faced with this situation.

INVERTED USAGE

Inverted usage can be applied when you want to emphasize the less frequently used functions and behaviors in the game. This creates a usage model that might reflect how the game would be used by people trying to find ways to exploit or intentionally crash the game for their own benefit. It also helps draw out defects that escaped earlier detection because of the very fact that the game is rarely, if ever, expected to be used in this way.

Calculating Inverted Usage

Inverted usage is calculated using a three-step process:

1. Calculate the reciprocal of each usage probability for a test parameter (combinatorial) or for all paths exiting a state (TFDs).
2. Sum the reciprocals.
3. Divide each reciprocal from Step 1 by the sum of the reciprocals calculated in Step 2. The result is the inverted probability for each individual usage value.

For example, say there are three values A, B, and C, with the usage 10%, 50%, and 40%, respectively.

Apply Step 1 of the inversion process to get reciprocal values of 10.0 for A $\left(\frac{1}{0.10}\right)$, 2.00 for B $\left(\frac{1}{0.5}\right)$, and 2.50 $\left(\frac{1}{0.40}\right)$ for C.

Add these reciprocals to get a sum of 14.5. The reciprocals ae divided by the sum to get the inverted values of 69.0% $\left(\frac{10}{14.5}\right)$ for A, 13.8% $\left(\frac{2}{14.5}\right)$ for B, and 17.2% $\left(\frac{2.5}{14.5}\right)$ for C. These can be rounded to 69%, 14%, and 17% for test generation purposes.

One characteristic of this process is that it inverts the proportions between each probability compared to its companions for a given set of usage values.

In the preceding example, B is used five times more frequently than A $\left(\frac{50}{10}\right)$ and 1.25 times more frequently than C $\left(\frac{50}{40}\right)$.

The relationship between inverted A and inverted B is $\frac{69\%}{13.8\%}$, which is 5.00.

Likewise, the relationship between inverted C and inverted B is 1.25 $\left(\frac{17.2\%}{13.8\%}\right)$.

For any case where there are only two usage values to invert, you can skip the math and simply reverse the usage of the two values in question. You will get the same result if you apply the full process, but why bother when you could use that time to do more testing?

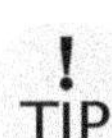

Inverting 0%
If an item has a 0% usage, then the first step in the inversion process will cause a divide by zero situation. Keep that from happening by adding 0.01% to each value before doing the three-step inversion calculation. This will keep the results accurate to one decimal place of precision in the results and maintain the relative proportions of usages in the same set.

Combinatorial Table Usage Inversion

Table 10.1 showed a set of usage probabilities for the *Halo: Reach* Look Sensitivity test values of 1, 3, and 10. Construct a table of inverted values starting with the Casual player profile. The three usage probabilities in that column are 10, 85, and 5. These are percentages, so the numerical values of these probabilities are 0.10, 0.85, and 0.05.

Apply Step 1 and calculate $\frac{1}{0.10} = 10$.

Do the same for $\frac{1}{0.85}$, which is 1.18, and $\frac{1}{0.05}$, which equals 20.

Add these numbers according to step 2. $10 + 1.176 + 20 = 31.176$. Finish with Step 3.

Dividing 10, which is the reciprocal of the usage probability for Look Sensitivity = 1, by 31.18, which is the sum of all three reciprocals, gives an inverted probability of 0.321. Because the numbers in the table are percentages, this gets entered as 32.1. Likewise, divide 1.18 by 31.18 to get the second inverted usage result 0.038, or 3.8%. Complete this column by dividing 20 by 31.18 to get 0.641 and enter 64.1 as the inverted usage for Look Sensitivity = 10.

Comparing the inverted usage values to the original ones confirms that the relative proportions of each usage value have also been inverted. Originally, the usage for Look Sensitivity = 1 was 10% versus 5% for Look Sensitivity = 10: a 2 to 1 ratio. In the inverted table, the Look Sensitivity = 10 value is 64.2—twice that of the 32.1% usage for Look Sensitivity = 1. You can examine the values for each parameter to confirm that this holds true for the other values within each column.

The complete inverted Look Sensitivity usage for all player profiles is provided in Table 10.14.

TABLE 10.14 Inverted usage percentages for the Look Sensitivity parameter

Look Sensitivity	Casual	Achiever	Explorer	Multiplayer
1	32.1	99.9	60.9	75.9
3	3.8	0.0	8.7	5.1
10	64.1	0.0	30.4	19.0
TOTAL	100	100	100	100

NOTE *The "normal" and inverted usage tables for all of the Halo: Reach Advanced Controls parameters are provided in a spreadsheet among the companion files for this chapter. There are separate worksheets for the normal and inverted usages. You can change the values on the normal usage sheet and the values on the inverted usage sheet will be calculated for you.*

All companion files for this book can be obtained by contacting the publisher at info@merclearning.com.

TFD Flow Usage Inversion

The TFD "Enter" and "Exit" flows present special cases you must address when inverting usages. Because these are really "test" operations versus "user" operations, the usage percentage for these flows should be preserved. They will keep the same value in the inverted usage set that you assigned to them originally. Table 10.15 shows the Unlock Item TFD's inverted Casual player usage table initialized with these fixed values.

TABLE 10.15 Inverted usage table initialized with "Enter" and "Exit" flow data

Flow	1	2	3	4	5	6	7	8	9	10	11	12	13	14
Casual	100					10								

Complete the table by performing the inversion calculation process for the flows leaving each state on the TFD. Go from state to state and fill in the table as you go along. Start at the top of the diagram with the "NoPointsItemLocked" state. Do inversion calculation for flows 3 and 14. Because these flows have the identical value of 50%, there is no need to do any math. The inverted result in this case is the same as the original. Put 50s in the table for these flows, as shown in Table 10.16.

TABLE 10.16 Fixed usage added for flows leaving "NoPointsItemLocked"

Flow	1	2	**3**	4	5	6	7	8	9	10	11	12	13	**14**
Casual	100		**50**			10								**50**

Moving clockwise around the diagram, do the inversion for flows 2 and 4 coming from "HavePointsItemUnlocked." There are only two values, so you can swap values without having to do a calculation. Table 10.17 shows the 90% inverted usage for flow 2 and the 10% inverted usage for flow 4 added to the table.

TABLE 10.17 Inverted usage added for flows leaving "HavePointsItemLocked"

Flow	1	**2**	3	**4**	5	6	7	8	9	10	11	12	13	14
Casual	100	**90**	50	**10**		10								50

The next state on your trip around the TFD is "HavePointsItemUnlocked." This is the state that has the "Exit" flow, which is already recorded as 10% in the inverted table. The trick here is to invert the other flows from this state while preserving the total usage of 100% when they are all added up, including the "Exit" flow. Have you figured out how to do this? For Step 1, calculate only the reciprocals of flows 5 (5%), 7 (75%), and 12 (10%). These would be 20, 1.33, and 10, respectively. The sum of the reciprocals (Step 2) is 31.33. Divide each reciprocal with the sum (Step 3) to get 0.638, 0.042, and 0.319. Because it has already been established that flow 6 (Exit) accounts for 10% of the usage probability total for "HavePointsItemUnlocked," then these other three flows must account for the remaining 90%. Multiply the inverted usages for flows 5, 7, and 12 by 0.9 (90%) to account for that. The final result for flow 5 is 0.574 (57.4%), for flow 7 is 0.038 (3.8%), and for flow 12 is 0.287 (28.7%). Table 10.18 shows these numbers included with the results for the other flows usages calculated so far.

TABLE 10.18 Inverted usage added for flows leaving "HavePointsItemUnlocked"

Flow	1	2	3	4	**5**	6	**7**	8	9	10	11	**12**	13	14
Casual	100	90	50	10	**57.4**	10	**3.8**					**28.7**		50

Go to the next state, which is "NoPointsHaveItem." This is another situation with only two flows to invert. Swap the usage values for flow 8 and flow 9. Table 10.19 shows flow 8 added to the table with a 90% inverted usage and flow 9 with a 10% inverted usage.

TABLE 10.19 Inverted usage added for flows leaving "NoPointsHaveItem"

Flow	1	2	3	4	5	6	7	**8**	**9**	10	11	12	13	14
Casual	100	90	50	10	57.4	10	3.8	**90**	**10**			28.7		50

"NoPointsItemUnlocked" is the last state to account for on the diagram. Three flows leave this state, so you have to do some calculations. Flow 10 has a 5% usage, so its reciprocal is 20. Flow 11 has a 90% usage. Its reciprocal is 1.11. Flow 13 has the same usage as flow 10 and, therefore, the same reciprocal of 20. Now do Step 2 and add up the reciprocals. 20 + 1.11 + 20 = 41.11. Find the inverted usage of each flow by dividing their reciprocals by this total.

[For flows 10 and 13, calculate $\frac{20}{41.11}$, which results in 0.486, or 48.6%.]

[Calculate flow 11's inverted usage as $\frac{1.11}{41.11}$, which is 0.027, or 2.7%.]

Enter these values to the table to get the completed version shown in Table 10.20.

TABLE 10.20 Completed table with inverted usage for "NoPointsItemUnlocked"

Flow	1	2	3	4	5	6	7	8	9	**10**	**11**	12	**13**	14
Casual	100	90	50	10	57.4	10	3.8	90	10	**48.6**	**2.7**	28.7	**48.6**	50

With these inverted percentages, you can produce TFD Cleanroom paths and test cases in the same way you did earlier from the normal usage probabilities.

One technique that makes it a little easier to keep track of the number ranges associated with each percentage is to add a Range column to the usage table. Table 10.21 shows how this looks for the Unlock Item TFD inverted usages. This column can be especially helpful when the flows from a state are scattered around, such as flows 3 and 12 coming from "NoPointsItemLocked."

Game players have tendencies and patterns of use that can be incorporated into game tests for the purpose of testing the game the way players play the game. The point of doing that is to find and remove the bugs that would show up when the game is played in those ways. If you are successful, those players will not find any bugs in your game. That is good for them and good for you.

When you sell millions of copies of your game, "rare" situations can show up dozens, if not hundreds, of times over the life of a title. Tests based on inverted usage profiles can emphasize and expose those rare defects in your game.

TABLE 10.21 Inverted Casual player usage and ranges for Unlock Item TFD

Flow	Casual Usage	Range
1	100	1-100
2	90	1-90
3	50	1-50
4	10	91-100
5	57.4	1-57
6	10	58-67
7	3.8	68-71
8	90	1-90
9	10	91-100
10	48.6	1-49
11	2.7	50-52
12	28.7	72-100
13	48.6	53-100
14	50	51-100

EXERCISES

1. Is it possible to have the same exact test case appear more than once in a Cleanroom test set? Explain.
2. Identify and list each pair of values that is missing from the Cleanroom combinatorial table in Table 10.12. Explain why they are not necessary and why they might not even be desirable in this application.
3. Give an example of "Real-Life Usage" data incorporated into gameplay from a game you have played recently.
4. Create a set of tables with the inverted Casual profile usage probabilities for each of the *Halo: Reach* Advanced Settings parameters.
5. Generate six Cleanroom combinatorial tests from the inverted usage tables you produced in Exercise 4. Use the same random number set that was used to generate the combinatorial tests shown in Table 10.12. Compare the new tests to the original ones.
6. Modify the TFD from Figure 10.4 to incorporate the inverted usages in Table 10.21. Round the usage values to the nearest whole percentage. Make sure the total probabilities of the flows exiting each state add up to 100. If not, adjust your rounded values accordingly.
7. Generate a path for the TFD you produced in Exercise 5. List the flows, actions, and states along your path using the same format shown earlier in this chapter. Compare the new path to the original one.

REFERENCES

Bartle, Richard. 1996. "Hearts, Clubs, Diamonds, Spades: Players Who Suit MUDs." Retrieved January 13, 2024. *https://mud.co.uk/richard/hcds.htm*.

Prowell, Stacy J., Trammell, Carmen J., Linger, Richard C., Poore, Jesse H. 1999. *Cleanroom Software Engineering: Technology and Process*. United Kingdom: Pearson Education.

CHAPTER 11

TEST TREES

Test trees can be used for three different purposes in game testing:

1. Test case trees document the hierarchical relationship between test cases and game features, elements, and functions.
2. Tree feature tests reflect the tree structures of features and functions designed into the game.
3. Test tree designs are used to develop tests that systematically cover specific game features, elements, or functions.

TEST CASE TREES

In this application of test trees, the tests have already been developed and documented. The tree is used each time the game team sends a new release to the testers. The test lead can determine which tests to execute based on which defect fixes or new abilities were introduced in the release. Such an organization could also reflect the way the game itself is structured.

Take, for example, a tree of tests for *Warhammer 40,000: Dawn of War*, a real-time strategy game for the PC. In this game, up to eight players can compete against one another or computer AI opponents. Players control and develop their own race of warriors, each of which has its own distinct military units, weapons, structures, and vehicles. Games are won according to various victory conditions, such as taking control of a location, defending a location for a given amount of time, or completely eliminating enemy forces.

At a high level, the *Dawn of War* tests can be organized into Game Options tests, User Interface tests, Game Mode tests, Race-specific tests, and Chat capability tests. The Option tests can be grouped into Graphics, Sound, or Controls options. The User Interface tests can be divided between the Game Screen UI and the in-game Camera Movement. Additionally, there are three major Game Modes: Campaign, Skirmish, and Multiplayer, and four Races that players can choose from: Chaos, Eldar, Orks, and Space Marines. The Chat capability is available when connected via LAN, Online, or Direct Link. Figure 11.1 shows these top two levels of organization arranged as a tree.

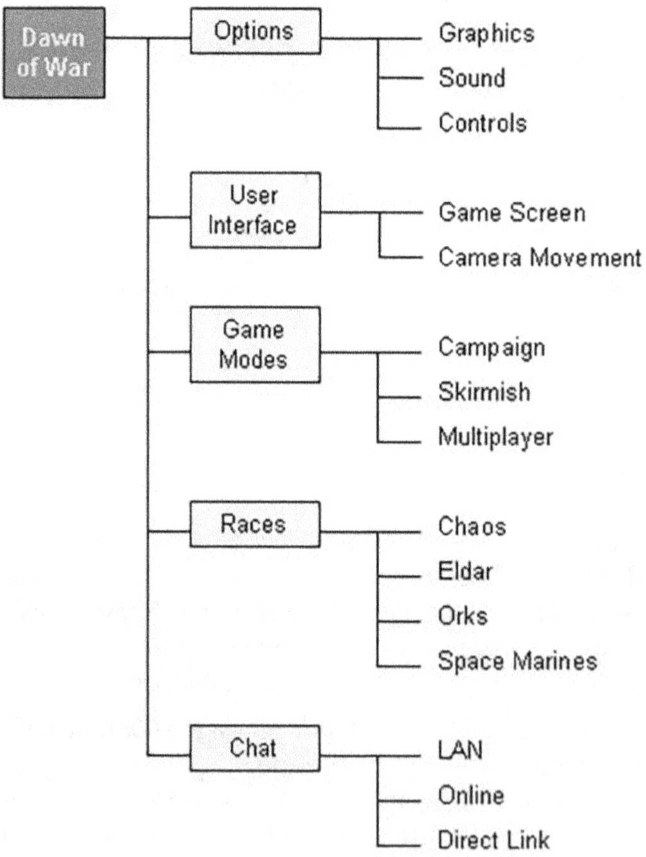

FIGURE 11.1 *Dawn of War* two-level test case tree

During game development, each bug fix can affect one or more areas of the game. With the test case tree, you can easily target which tests to run by finding them under the tree nodes related to the parts of the game affected by the new code. Some fixes could have to be re-checked at a high level, such as a change in the Chat editor font that applies to all uses of chat. Other fixes might be more specific, such as a change in the way Chat text is passed to the online chat server.

It is also possible to define the tree in finer detail to make a more precise selection of tests. For example, the Skirmish Game Modes tests could be further organized by which map is used, how many players are active in the match, which race is chosen by the player, what game options are selected, and which win conditions are applied. Figure 11.2 shows the further breakdown of the Skirmish branch.

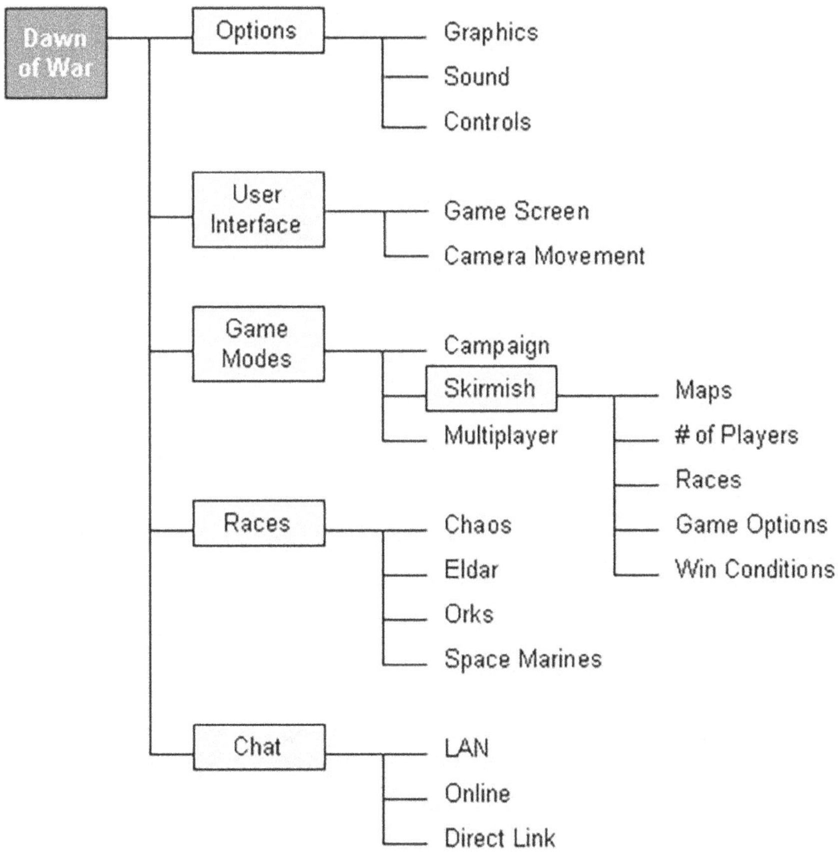

FIGURE 11.2 Skirmish Game Mode test case sub-tree added

Revealing the additional details of the Skirmish mode is important because it exposes another set of tests that should be run if changes are made to any game assets or functions that are specific to one or more of the Races. Whether your tests are stored in a regular directory system, configuration management repository, or test management tool, you can organize them to match the tree hierarchy of the game's functions. This is an efficient way to find the tests you want to run once you map them to the code changes in each release you test.

TREE FEATURE TESTS

A second application of test trees is used to reflect the actual tree structure of features implemented in the game. *Dawn of War* has such structures for the technology trees of each race. These trees define the dependency rules for which units, vehicles, structures, and abilities can be generated. For example, before the Eldars can produce Howling Banshee units, they must first construct an Aspect Portal and upgrade the structure with the Howling Banshee Aspect Stone. Other units can be produced immediately, such as the Rangers. These trees can be quite complex, with dependencies between multiple structures, upgrades, and research items. Test these trees by following the various paths to successfully construct each item. Check that attempted "shortcuts" will not produce the intended result, such as trying to produce Warp Spider units without the Warp Spider Aspect Stone. Be thorough and examine all the ways this might be attempted, such as from a menu, command line, or by clicking an icon. Figure 11.3 shows the Aspect Portal tech tree for the Eldar race.

Another example of this type of tree is the job or skill tree typically defined for RPG games, such as the *Final Fantasy* or *Dragon Age* series. Characters might be required to develop skills up to a certain level before new skills or abilities become available. In some cases, the skill and role choices are dictated by the choice of character race, occupation, or faction. Each successive choice will perhaps narrow the options available for the remaining choices. For these kinds of trees, think of the string of lights on a Christmas tree. If one light is faulty, the remaining connected lights will also be off. In this case, some Classes or Backgrounds will not be available if the preconditions (required combinations of previous choices) are not met. As a tester, you will want to test each possible result by leaving out the necessary preconditions, one at a time. In addition, you will test for the case where all of the necessary conditions are met.

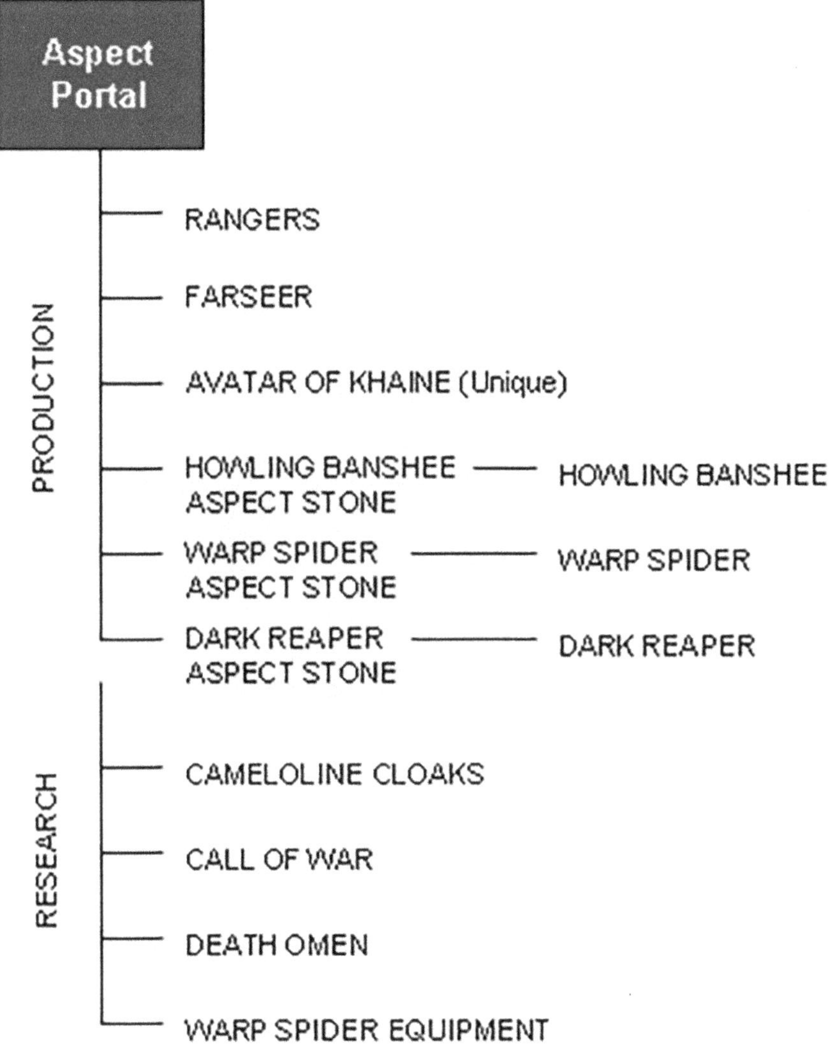

FIGURE 11.3 *Dawn of War* technology tree for Eldar Aspect Portal

See Figure 11.4 for an example of how this is portrayed in the game screen for a male dwarf in *Dragon Age: Origins*. Selecting the "male" gender reveals the Race choices that are available. In this case, any of the three races—Human, Elf, or Dwarf—can be chosen. Choosing the Dwarf race limits your choice of class to warrior or rogue.

FIGURE 11.4 Male Dwarf character generation in *Dragon Age: Origins*

Selecting the rogue class for the Dwarf male limits you to two of the six background possibilities available: commoner or noble.

When the character tree is not already drawn for you on the game screen, you may have to construct your own. As you progress through the tree, check that the allowed roles and skills are available at the end of the tree, and check that choices that should be unavailable are blocked along the way.

Job trees are another construct built into many popular games. Take a moment to look at Figure 11.5 which shows a tree diagram for the mage jobs available to Hume (human) characters in *Final Fantasy Tactics A2: Grimoire of the Rift* (*FFTA2*).

Before he can take on the role of an illusionist, the player's character must master two white mage abilities and four black mage abilities. The diagram also shows that the character must master one white mage ability before he can even take on the role of black mage to begin mastering black mage abilities.

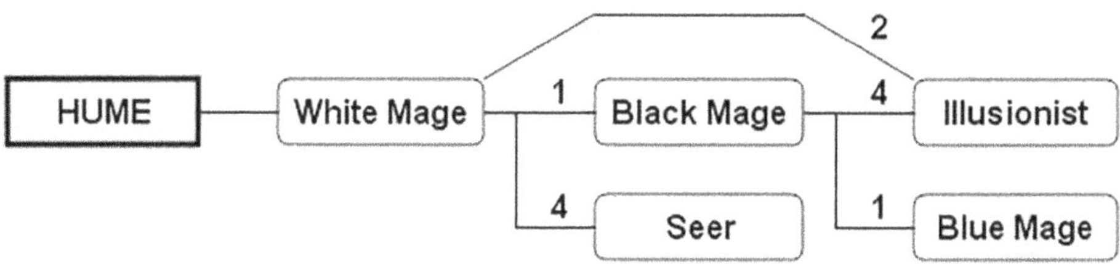

FIGURE 11.5 Hume Mage job tree for *FFTA2*

Define the tree feature tests for a particular job by providing the test values for each of the nodes along the tree branches. You can keep the tests to a minimum by checking only the "edges" of the values that determine whether a skill, job, or ability is unlocked. For example, there are no tests for white mage mastery = 2 nor black mage mastery = 1 or 2. It is sufficient to know that the illusionist job is still not unlocked when only three black mage abilities are mastered in test case 2, and then establish that it is properly unlocked when black mage proficiency reaches 4. Table 11.1 shows the test cases that should result for the tree in Figure 11.5.

TABLE 11.1 Hume Mage illusionist job tree tests

Test	White Mage	Black Mage	Illusionist Enabled
1	1	4	No
2	2	3	No
3	2	4	Yes

Trees can also define and limit which areas or locations within a game can be accessed by the player at any given time. Figure 11.6 shows the progression of battlegrounds that become available to players of the mobile RPG *Battleheart*. Check marks indicate locations conquered, skulls represent unconquered locations that are available to the player, and padlocks indicate locations that are not accessible.

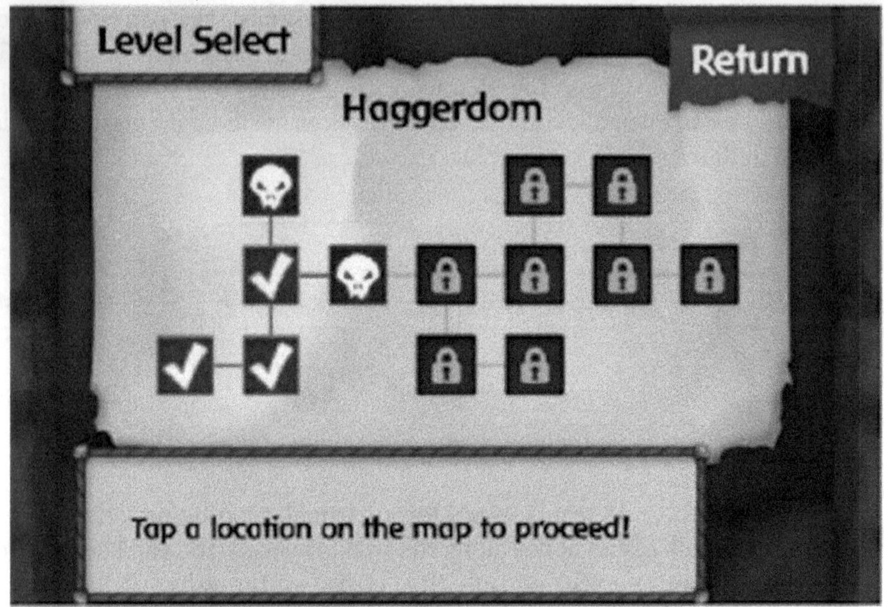

FIGURE 11.6 *Battleheart* battleground selection tree

Many other feature trees exist across a wide variety of games. Here is a list of some places you might find them:

- technology trees
- progressing through tournament brackets
- game option menu structures
- adding or upgrading spells, abilities, or superpowers
- increasing complexity and types of skills needed to craft or cook items
- earning new ranks, titles, trophies, or medals
- unlocking codes, upgrades, or power-ups
- unlocking new maps, battlegrounds, environments, or race courses
- unlocking better cars, outfits, or opponents

One situation that is especially interesting to test arises when different menu trees or tree paths can affect the same value, ability, or game element. Such values should be set and checked by all the possible paths provided by the game. For example, the *Dawn of War* game options that are set in Skirmish mode also become the values used in the multiplayer, LAN, online, and direct host modes.

TEST TREE DESIGNS

This chapter has so far dealt with intentionally organized paths and patterns built into the game for the player's progression, improvement, or advancement. At the other end of the spectrum, some game systems can seem incredibly chaotic. Take card battle games as an example. In these types of games, players take turns revealing cards from a deck they have assembled, which must also meet certain specifications imposed by the game rules. Winning the card battle usually involves eliminating the opponent, his creatures (army, players, etc.), or both. A card will perhaps have a special behavior that is defined by the type of card or additional instructions printed on the card. Some cards can affect other players or cards. There are cards for offensive and defensive purposes. Hundreds of different cards can potentially interact and affect each other in unexpected or undesirable ways. More cards become available over time, creating new and sometimes unexpected interactions. In the design of these games, the seeming chaos and unexpected interactions are a highlight of the gameplay. However, all these interactions must be tested to ensure that the cards and card combinations are behaving in ways that seem true to the design of the game. You can create a test tree design to define a set of tests for special card capabilities.

Magic: The Gathering® is a super popular collectible card game available in both physical and video game form. A computer version, *Magic: The Gathering—Duels of the Planeswalkers*™, includes a Challenge mode in which the player has a single turn to defeat their opponent, who seems to have an insurmountable advantage. Only a specific combination of cards, played in the correct order and put to proper use, will result in victory.

In a *Magic: The Gathering* duel, cards are played that provide energy ("mana") of various types (red, blue, green, white, black, or colorless) which are used to summon creatures or power spells and abilities. Assigning the right types of mana to each card played is key to enabling subsequent spells or abilities to be played in the same turn.

Challenge 5 has you playing against Liliana with only two life points left to your name. You have to determine how to damage Liliana for 13 points or more in one turn if you are going to win. You have six cards in your hand, plus five creatures in play, and six mana—three green (G) and three black (B)—on the table to power your cards. The cards and their costs are as follows:

Overrun – 2GGG

Elvish Champion – 1GG

Imperious Perfect – 2G

Elvish Warrior – GG

Elvish Eulogist – G

Eyeblight's Ending – 2B

You can organize the possible sequences of played cards into a tree structure, accounting for all of the choices remaining after each subsequent play. This will reveal which sequences (paths) will lead to defeat and which (at least one) should result in success. As an alternative to laying out trees in a graphical format, you can use a spreadsheet to organize and visualize the structure and relationships between game options, player choices, and the results for each branch.

Start defining the paths, using the most expensive card to play. This reduces the amount of mana left with which to play subsequent cards so it should be one of the simpler branches in the tree/table. In this scenario, the Overrun card has the highest cost of 2GGG, which means the player has to pay three green-colored mana and two additional mana of any color (green included) to play this spell. Once that has been played, only one black mana is left. This is not enough to pay for any of the remaining cards. Attacking after playing this card and applying any of the effects available to your other cards in play will not result in a victory, so this is a losing path. See Table 11.2 for the representation of this branch of the test tree in a tabular (spreadsheet) format. Because each path is a test case, they should be numbered for easy reference.

TABLE 11.2 Overrun card branch for the Liliana Challenge

Test	1st Card	Payment	2nd Card	Payment	Result
1	Overrun (2GGG)	BGGG	N/A	N/A	Lose

The next most restrictive card is Elvish Champion, which costs 1GG—two green mana and one other mana of any color. Some rows for this branch end with no 2nd card played, because a player could choose to complete their turn even if they have not used up all of their available mana. The "1" means any type of available mana could be paid, so there needs to be a separate sub-branch for when green mana is used for the "1" and another for when black mana is used. These choices will provide different constraints on which cards could possibly be played with the remaining mana available. A key to creating a thorough table is to recognize that different action "payment" combinations for each node in the tree must all be accounted for. Additional sub-branches are necessary to represent the various options for playing a second card. See Table 11.3 for the Elvish Champion branch of the test tree. Take note that in the last branch of this table, you will still have two black mana available after paying a total of BGGG, but no further move is possible because there are no cards in your hand which can be played for that cost.

TABLE 11.3 Elvish Champion card branch for the Liliana Challenge

Test	1st Card	Payment	2nd Card	Payment	Result
2	Elvish Champion (1GG)	GGG	None	N/A	Lose
3			Eyeblight's Ending (1BB)	BBB	Lose
4		BGG	None	N/A	Lose
5			Eyeblight's Ending (1BB)	BBG	Lose
6			Imperious Perfect (2G)	BBG	Lose
7			Elvish Eulogist (G)	G	Lose

NOTE *Sometimes it is a clue to the player when all of the resources provided, such as mana, must be used to solve a problem or to win a challenge. While this is not a bug per se, a tester should point this out when it is to the detriment of the game, such as when there is only one way to spend the resources, and that is also the winning move.*

Next is the branch for playing the Imperious Perfect card first. This card costs 2G to play: one green mana plus two more of any color. With the mana available for the challenge, the possible costs for this card are GGG, BGG, and BBG. Here you encounter a greater number of restrictions as certain colors of mana are used up, such as the inability to play the Elvish Champion card (1GG) when paying BGG for the Imperious Perfect. Table 11.4 maps out the card play sequences available when starting your turn with the Imperious Perfect card.

TABLE 11.4 Imperious Perfect card branch for the Liliana Challenge

Test	1st Card	Payment	2nd Card	Payment	Result
8	Imperious Perfect (2G)	GGG	None	N/A	Lose
9			Eyeblight's Ending (2B)	BBB	Lose
10		BGG	None	N/A	Lose
11			Eyeblight's Ending (2B)	BBG	Lose
12			Elvish Eulogist (G)	G	Lose
13		BBG	None	N/A	Lose
14			Eyeblight's Ending (2B)	BGG	Lose
15			Elvish Champion (1GG)	BGG	Lose
16			Elvish Warrior (GG)	GG	Lose
17			Elvish Eulogist (G)	G	Lose

Continue by building out the Elvish Warrior card branch. This card costs only two mana (GG), so there might be opportunities for more than two cards to be played, depending on how the mana is spent. Once the Elvish Warrior card is followed by Elvish Eulogist, you choose not to play any more cards, but are still capable of paying for Eyeblight's Ending with the threeblack mana that remain. Additional columns need to be added to the test tree table to accommodate the possible "3rd card nodes" of the test tree. Table 11.5 shows the table for the Elvish Warrior card, which provides not just one, but TWO winning branches.

TABLE 11.5 Elvish Warrior card branch for the Liliana Challenge

Test	1st Card	Payment	2nd Card	Payment	3rd Card	Payment	Result
18	Elvish Warrior (GG)	GG	None	N/A	N/A	N/A	Lose
19			Imperious Perfect (2G)	BBG	N/A	N/A	Lose
20			Elvish Eulogist (G)	G	None	N/A	Lose
21					Eyeblight's Ending (2B)	BBB	**WIN**
22			Eyeblight's Ending (2B)	BBG	N/A	N/A	Lose
23				BBB	None	N/A	Lose
24					Elvish Eulogist (G)	G	**WIN**

To win this challenge, you have to do more than just put the proper cards into play. When Eyeblight's Ending is played, you need to target (destroy) the opponent's Nightmare card, which is her only creature with the Flying ability. After the Elvish Eulogist is played, you need to "tap" the Immaculate Magistrate card—which was already in play at the start of the challenge—to increase the amount of damage that will be done by the Elven Riders card. With the Nightmare out of the way, your Elven Riders cannot be blocked, so when you attack, you deliver 13 points of unblocked damage to defeat Liliana.

At this point, two more main branches remain to be explored. The Elvish Eulogist card is the last one that requires a green mana cost, as shown in Table 11.6. Two more winning sequences are revealed.

TABLE 11.6 Elvish Eulogist card branch for the Liliana Challenge

Test	1st Card	Payment	2nd Card	Payment	3rd Card	Payment	Result
25	Elvish Eulogist (G)	G	None	N/A	N/A	N/A	Lose
26			Elvish Warrior (GG)	GG	N/A	N/A	Lose
27					Eyeblight's Ending (2B)	BBB	**WIN**
28			Imperious Perfect (2G)	BGG	N/A	N/A	Lose
29				BBG	N/A	N/A	Lose
30			Eyeblight's Ending (2B)	BGG	N/A	N/A	Lose
31				BBG	N/A	N/A	Lose
32				BBB	N/A	N/A	Lose
33					Elvish Warrior (GG)	GG	**WIN**

The last tree to construct starts when Eyeblight's Ending is played first. This opening move produces the largest table for this challenge. This time the player is prevented from accessing the two winning branches in this portion of the tree because the game will not let you pay three black mana (BBB) for the Eyeblight's Ending card. Instead, it automatically pays two black mana and one green mana (BBG).

As you play through the card sequences in this final portion of the tree, you should notice that when you play the Eyeblight's Ending card as your first card, the game automatically chooses to pay with BBG. In the version of the game that serves as the basis for this example, there seems to be no mechanism in the game for the player to explicitly which lands (mana) are used to pay for each card as it is played, so the nodes on the test branch that require the player to pay BGG or BBG cannot be executed, preventing access to two winning branches in tests 44 and 46. Woe to the players who open with Eyeblight's Ending, for they may find themselves quite frustrated.

There are also a number of losing plays which are blocked for the same reason, denying players the opportunity to explore those possibilities. For example, when playing Eyeblight's Ending after the Elvish Eulogist in tests 32 and 33, the game uses BBB and does not give the option to pay GBB as an alternative. This blocks tests 30 and 31. Looking back over the entire test tree, a total of 21 branches—nearly half of the tree—cannot be played:

- 2, 3 – cannot pay GGG for Elvish Champion
- 8, 9 – cannot pay GGG for Imperious Perfect
- 13, 14, 15, 16, 17 – cannot pay BBG for Imperious Perfect
- 28 – cannot pay BGG for Imperious Perfect
- 30 – cannot pay BGG for Eyeblight's Ending
- 31 – cannot pay BBG for Eyeblight's Ending
- 34, 35 – cannot pay BGG for Eyeblight's Ending
- 40 through 46 – cannot play BBB for Eyeblight's Ending

These tests should still remain in your test design, but they could be grayed out or designated as "blocked" until either the AI is updated to use

TABLE 11.7 Eyeblight's Ending card branch for the Liliana Challenge

Test	1st Card	Payment	2nd Card	Payment	3rd Card	Payment	Result
34	Eyeblight's Ending (2B)	BGG	None	N/A	N/A	N/A	Lose
35			Elvish Eulogist (G)	G	N/A	N/A	Lose
36		BBG	None	N/A	N/A	N/A	Lose
37			Elvish Champion (1GG)	BGG	N/A	N/A	Lose
38			Imperious Perfect (2G)	BGG	N/A	N/A	Lose
39			Elvish Eulogist (G)	G	N/A	N/A	Lose
40		BBB	None	N/A	N/A	N/A	Lose
41			Elvish Champion (1GG)	GGG	N/A	N/A	Lose
42			Imperious Perfect (2G)	GGG	N/A	N/A	Lose
43			Elvish Warrior (GG)	GG	N/A	N/A	Lose
44					Elvish Eulogist (G)	G	**WIN***
45			Elvish Eulogist (G)	G	N/A	N/A	Lose
46					Elvish Warrior (GG)	GG	**WIN***

the mana differently or the game is updated to provide a way for players to control how mana is spent for each card.

Tree structures are useful for organizing test cases so that the proper set of tests can easily be selected for a given set of changes to the game. Each downstream node represents a set of tests with a more specific purpose and scope than the nodes above it. Tests can also reflect tree-like relationships that exist between game functions and elements. The behavior of these structures is tested by exercising the values along the various possible paths from the start of the tree, through each branch, and ending when there are no more moves, decisions, or choices to make.

Finally, test trees can be designed to improve understanding of a complex game feature or problem to be solved, and to bring order to a seemingly chaotic or complex function. This is especially relevant when you need to explore the interaction of game rules, options, elements, and functions. A well-formed tree progressively decomposes the feature until the end nodes are reached, defining the specific actions to perform during testing. Do not forget to "prune" any branches that are not possible due to game limitations.

EXERCISES

1. From the test case tree in Figure 11.2, which test branch(es) should you re-run for a new release of the game that fixes a bug with the sound effect for the Orks "Big Shoota" weapon?
2. There are actually four multiplayer game modes in *Dawn of War*: LAN, Online, Direct Host, and Direct Join. Furthermore, the same choices are available in Skirmish mode—Maps, # of Players, Race, Game Options, and Win Conditions—apply to the LAN and Direct Host multiplayer modes. Describe how you would update the test case tree in Figure 11.2 to include these additional choices.
3. Draw a test tree representing the following relationships between lessons and items in *The School of Wizardry*.
   a. You can find a Disarming Spell by doing the "Discover that you possess wizard powers as well" lesson.
   b. The "Get a wand of your own" lesson requires the Disarming Spell and can also give you the Flashlight Charm.

   c. The "Go to your first day of nonwizard school" lesson requires one Flashlight Charm and can provide an Impeding Charm.
   d. "Your first magic lesson with Uncle Mortimer—levitate an object" lesson requires one Impeding Charm and can give you a Confusion Spell.
   e. The "Study the History of Magic" lesson requires two Confusion Spells and can provide a Cast Flame Charm.
   f. The "Study potions with your uncle" lesson requires two Disarming Spells.
   g. The "Escape the neve-rending path in the Mystical Forest" lesson requires three Cast Flame Charms.
   h. The "Make it back home safely…" lesson requires five Cast Flame Charms.
4. Copy and edit the spreadsheet included in the companion files for this chapter to update the "Liliana Challenge" tree table to show how it should look if the Imperious Perfect card were to cost 1GG to play instead of 2G.

NOTE *All companion files for this book can be obtained by contacting the publisher at info@merclearning.com.*

CHAPTER 12

DEFECT TRIGGERS

We first discussed Orthogonal Defect Classification (ODC) in Chapter 4, "How Bugs Happen." That system includes a set of defect triggers to categorize the way defects are caused to appear. These same triggers can be used to classify tests as well as defects. Test suites that do not account for each of the triggers will be incapable of revealing all of the defects in the game.

OPERATING REGIONS

Game operation can be broken down into three stages: Game Start, In-Game, and Post-Game. These regions do not just apply to the game as a whole. They can also be mapped to discrete experiences within the game, such as new missions, seasons, or levels. There is also a Pre-Game region in which the game environment (hardware, operating system, and so on) is operational but the game has not been started. Figure 12.1 shows the relationship of these operating regions.

Pre-Game Operating Region

The *Pre-Game operating region* represents the period that precedes the use of the game. For consoles, this would be the time prior to inserting the game disk, or while browsing in the lobby to choose which game to play. On PCs and mobile phones, this is the period in time before you launch the game app. Cartridge-based handhelds will also have an operational mode that is used prior to the insertion of the game cartridge. In each of these cases, the user can change settings and do things with the device that will potentially impact the subsequent operation of your game.

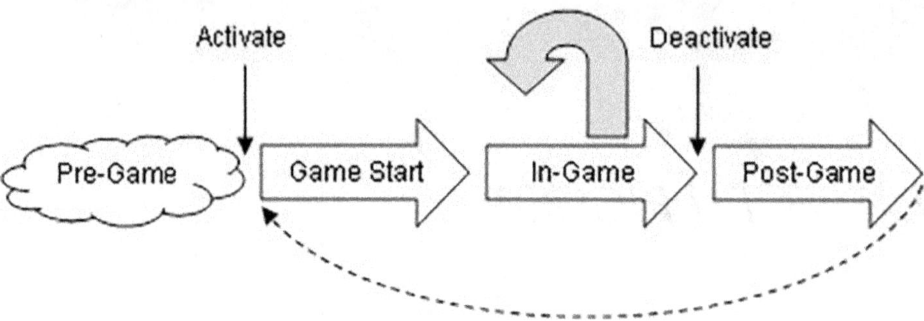

FIGURE 12.1 Game software operating regions

Game Start Operating Region

The *Game Start operating region* accounts for operations that are performed from the time the player starts the game until the time the game is actually ready to be played. Some activities that take place during this time can be interrupted, such as cinematic sequences that provide an introduction or highlights of the game's features. Other activities, such as a screen displaying the "loading" progress, cannot be accelerated or interrupted. The game software also performs activities that are essential to the proper operation of the game but are not visible to the player. At the very end of this process, the game could be in a "ready" state, during which it is waiting for the player to hit a button or key in order to enter the game.

In-Game Operating Region

The *In-Game operating region* includes all of the actions you could possibly make when playing the game. Some functions can be performed only once during the course of the game, whereas others can be repeated throughout the game. There are also functions that depend on the player meeting some condition before they can occur. Games that incorporate NPCs also manage and control these resources during this operating period.

Post-Game Operating Region

The player can end the game or a gaming session a number of ways. Quitting without saving requires less processing than when saving. The player is often given the opportunity to save character data and progress before the game terminates itself. Games played on portable devices can be ended by turning off the device. If the device's "off" switch is under software control, the game software can perform save and shutdown operations prior to killing power.

Story-based games often treat the user to a final cinematic sequence and roll credits when users reach the end of the story. Some games unlock new experiences for the player who reaches the end so he can continue to enjoy the game when going back through it a second time. This could activate code that is not exercised at all until the first time the game is completed.

THE TRIGGERS

Six defect triggers span the four game operating regions. These triggers describe ways to cause distinct categories of game defects to show up during testing. Together, these triggers account for all of the possible defects that can occur.

The Configuration Trigger

Some game configuration takes place in the Pre-Game region, prior to running the game. This includes device or environment settings that are established before running the game, such as game platform software versions. Date and time, screen resolution, system audio volume, operating system version, patches, and language settings are all examples of *Configuration triggers*. Figure 12.2 shows the many video configuration settings available for *Mass Effect 2* on the PC.

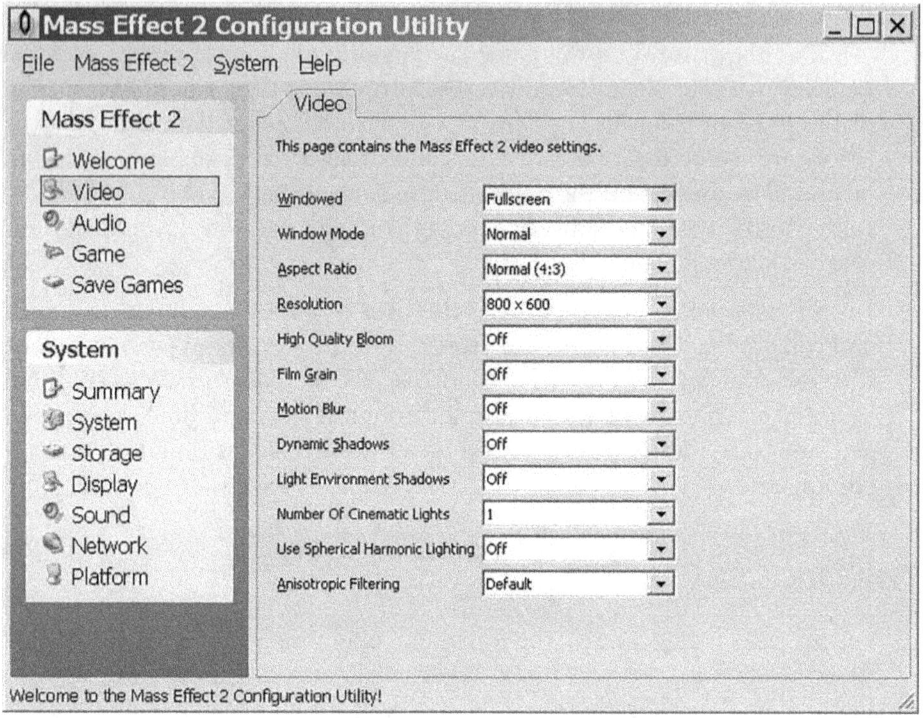

FIGURE 12.2 *Mass Effect 2* PC Video Configuration settings

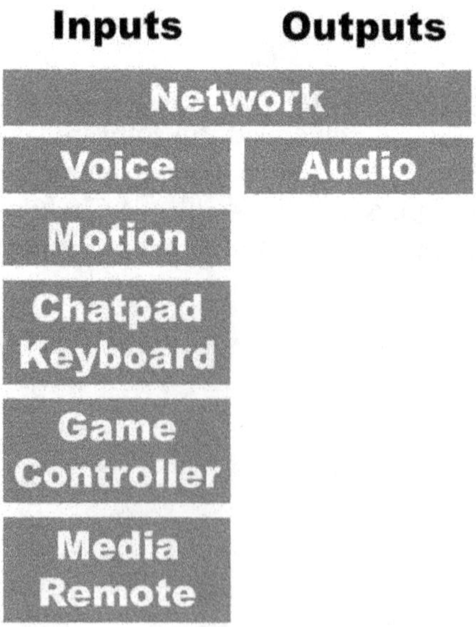

FIGURE 12.3 Xbox One I/O Interfaces

Configuration also involves external devices that can be used with the game platform. Game controllers, keyboards, mice, speakers, monitors, and network connections are all parts of the test configuration. These devices typically connect to the game console's I/O (Input/Output) ports through various connectors or wireless receivers. An Xbox One interface diagram is shown in Figure 12.3.

The nature of these external devices is that they each have their own settings, properties (for example, version), and physical connection. Even game controllers can have additional devices and modes of operation. A Nintendo Wii U console can be used in combination with a Wii Remote™, Wii U Game Pad, Nunchuck, Balance Board, WiiWheel™, or MotionPlus™ accessory. An Xbox One user can play from the standard game controller or use a Chatpad and headset combination.

Disconnecting one or more devices during gameplay is a type of configuration change. Unfortunately for developers, the game software is unable to do anything to prevent the user from connecting or disconnecting external devices during gameplay, or from changing settings or software versions on the game platform. Configuration changes can occur in any of the game software operating regions.

Connecting a device could be done to correct an accidental disconnection ("The dog kicked the wireless router plug out of the wall!"), replace batteries, change out a faulty device, or add a new capability, such as a headset for voice control. These scenarios should be anticipated by the game design and incorporated into the game tests.

Configuration possibilities should not be excluded from testing just because your initial response is "Why would anyone ever do that?" When you have this kind of reaction, recognize that you should test that area vigorously. It is likely that other people would have reacted similarly and did not bother to find out what would happen in that case.

Configuration failures might show up immediately as a result of the configuration operation or at some later time when a game operation relies on the new configuration. Seemingly unrelated capabilities might also fail as a side-effect of a configuration change.

The Startup Trigger

The *Startup trigger* is utilized by attempting operations while some game function is in the process of starting up, or immediately after that while code values and states are in their initial conditions. This could be a highly noticeable activity, such as a "Loading please wait…" screen, or a series of messages that are updating you of the progress being made during the startup process. Other events may happen entirely "behind the scenes," such as waiting for the game to load graphic content for a room you just entered, or being unable to proceed until a remote server authenticates your in-game identity.

Particular code vulnerabilities exist during the startup period. These do not present themselves at any other time in the game. Code variables are being initialized. Graphics information is being loaded, buffered, and rendered. Information is read from and written to a server or the local device's memory.

As an example, here is a summary of the events that take place in the Unreal Developer Kit (UDK) to start up a new level:

5. The GameInfo's `InitGame()` event is called.
6. The GameInfo's `SetGrammar()` event is called.
7. All Actors' `PreBeginPlay()` events are called.
8. All Actors' zones are initialized.
9. All Actors' `PhysicsVolumes` are initialized.
10. All Actors with `bScriptInitialized=false` have their `PostBeginPlay()` functions called.
11. The `SetInitialState()` function is called on all actors with `bScriptInitialized=false`.
12. Actors are "Attached" based on their `AttachTag`, `bShouldBaseOnStartup`, `Physics`, and world collision settings.

Startup defects are triggered by operations that take place during the Game Start period. These operations can be user-initiated or can be caused by the game platform. Interrupting any part of this sequence could mean that some essential operation will not complete its work or might not get to run at all. The Startup trigger accounts for bugs that will show up only as a result of the initial conditions that result from the game's initialization and startup processes. That means that defects that occur the very first time you use a game capability, such as a new map, item, power-up, or spell, should also be classified as Startup defects.

The Exception Trigger

Special portions of the game code are exercised by the *Exception trigger*. Exception handling in a game is normally recognized by the player. Audio "bonks" or alert boxes are common ways in which an in-game problem is communicated. Some exceptions are under the control of the game, such as restricting user input choices.

Other exceptions are caused by external conditions that are not under the control of the game software, such as network connection problems. Figure 12.4 shows the special alert you get when trying to play *Godville* if the mobile device is not connected to the Internet. Exceptions can occur in any of the game operating regions.

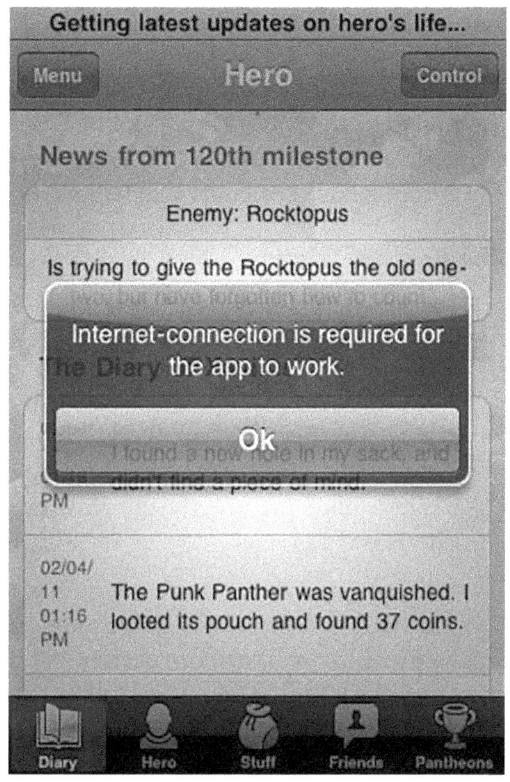

FIGURE 12.4 *Godville* connection exception alert

The Stress Trigger

The *Stress trigger* tests the game under extreme conditions. These could be conditions imposed on either hardware or software resources. Memory, screen resolution, disk space, file size, and network speed are all examples of game conditions that could be stressed by users or through testing. Simply reaching a limit does not constitute a stress condition. Once stressed, the resource must be used or operated in some way for the stress behavior to reveal itself.

The Normal Trigger

Normal game operation takes place in the In-Game operating region. This refers to using the game apart from any stress, configuration, or exception conditions, similar to the way you would script a demo or describe in the user manual how the game should be played. The "normal" code is distinct from the code that handles the exceptions, the code that processes configuration changes, and the code that takes over under stressful conditions.

Most of the testing that is done uses *Normal triggers*. That is fine, because that is how the game will be used the vast majority of the time; testing is not just about finding defects, it also demonstrates that the game functions the way it is supposed to. Testing that uses Normal triggers almost exclusively, however, is only training the code to follow a script. It will not detect many user faults that will occur in real-life situations.

The Restart Trigger

The *Restart trigger* classifies a failure that occurs as a result of quitting, ending the game, turning off the game device, ejecting the game disk, or terminating the game's operation in any other way. Some games are nice about this and prompt you to save vital information before allowing you to exit a game scenario, mission, level, or ongoing battle. When ending the game, some information needs to be remembered and some forgotten. If either is done incorrectly, the player can gain an advantage or lose precious progress.

You can "restart" under various conditions, such as when a game takes you back to a level selection screen, or when you load a saved game after failing. Be sure to explore each of the different restart methods available for the game you are testing and follow through by playing the game for a while, after the reloads, in order to detect any undesirable effects of the restart.

CLASSIFYING DEFECTS

You do not have to wait for your next project to start using defect triggers in your tests. Use keywords to help classify new or existing tests and defects. With that information, you can identify where the defects are coming from and what testing is missing. Not surprisingly, many bugs that get released belong to defect triggers that received little attention, if any, during game testing.

! TIP

Remain Vigilant!

When defects of a certain trigger start to show up, that is your cue to revise the tests to use that trigger even more often.

Table 12.1 provides a list of keywords you can use to classify your defects and tests according to each of the six defect triggers.

TABLE 12.1 Defect trigger keywords

Trigger	Keywords
Configuration	configure, model, type, version, environment, add, remove, setup
Startup	startup, initial, first, uninitialized, creation, boot, warm-up, wake-up, loading, transition
Exception	exception, error, violation, exceeds, NULL, unexpected, recover, prevented, blocked, prohibited, unavailable
Stress	stress, load, rate, slowest, fastest, low, high, speed, capacity, limit, long, short, few, many, multiple, empty, full
Normal	normal, typical, common, usual, expected, planned, basic, default, out-of-the-box, allowed, available
Restart	restart, reset, reload, cleanup, eject, power down, ctrl-alt-del, quit

Following are some examples taken from a list of bug fixes in version 1.2.214 of *The Elder Scrolls IV: Oblivion* (Nexus Mods 2008). Remember, missing capabilities are defects as well as game functionalities that do not work properly.

"Fixed issue where stolen items would lose their stolen status if the player character was female."

Because the character gender is established prior to entering the game world, this should be identified as a Configuration issue.

"Fixed a crash with stealing an object, exiting and immediately re-entering an interior."

In this situation, the problem only manifests when the interior location is re-entered. This should be considered a Restart fault.

"Fixed memory leak with sitting in a chair multiple times."

The Stress keyword "multiple" is used here, and it is in reference to a problem that occurs when trying to sit in a chair over and over again, so this is a Stress defect.

"Player can no longer fast travel when paralyzed."

Fast travel is a map-based way of traveling between landmarks in the *Elder Scrolls* world. Being paralyzed is not a configuration because it is a situation that occurs after the character has been configured and is active in the game world (the In-Game operating region). Neither is the fast travel ability a result of a particular configuration. This is simply a Normal defect.

"Fixed issue where lock/unlocked states on doors would occasionally be stored incorrectly in a saved game."

In this context, locking and unlocking is the door's "life cycle" in the game world. Loading the saved game data "restarts" the player's character and the state of all of the game assets. This restart loses track of the door's proper state. The defect has been revealed by a Restart trigger.

"Improved LOD visual quality for landscape."

Problems do not just have to happen in game logic to be considered bugs. Here, the level of detail of the rendered landscape was improved. The solution is not related to a particular condition or configuration of the game. This is a Normal defect.

"Fixed issue with LOD not loading in properly when entering/exiting world spaces."

Game maps can also have a life cycle: Start Map–Use Map–Change/Restart Map and so on. The "world space" is rendered with an undesirable level of detail as a result of the map restart. Therefore, this defect is triggered by that Restart.

"Pickup sound effects no longer play during the loading screen."

Yet another "life cycle" reveals itself here. The cycle of examining an item: select the item to examine, pick it up, examine it, then keep it or drop it. Because the problem is tied to the "loading" screen, map this to the Startup trigger.

"Fixed an occasional crash with NPCs who were not loaded going into combat."

This is a case where the game is referencing one or more "unavailable" resources. This is an Exception trigger defect.

"In Light the Dragonfires, fixed issue where improper journal would appear if you closed the Oblivion gate."

Do not confuse a game mission or quest with "configuration." Think of a quest as a feature or function of the game. Even though this bug only appears in a particular quest, the problem was in the In-Game operating region and not dependent on any configuration. It is another Normal trigger defect.

Sometimes you will come across defects that perhaps seem to belong to more than one trigger category. An example might be the case where an exception is not handled properly during startup. What you must resolve is which situation was primarily responsible for triggering the defect. If the situation is considered only an "exception" during startup, then it is the exception that is triggering the fault. The rationale is that there is a particular piece of code that should exist to handle the exception during startup. The exception condition causes that code to execute and finds that it is missing or faulty. Conversely, if the handling of that exception is common throughout the game, and it fails to operate properly only during startup, then it is the fact that you tried it at startup that triggered the exception code not to run or to run improperly. Your responsibility as a tester is to test the handling of this exception in all operating regions of the game in order to help make this kind of determination.

DEFECT TRIGGERS AND TEST DESIGNS

Each element in the test design represents one of the defect triggers. Using one or more test design techniques will not by itself guarantee that all of the defect triggers will be sufficiently represented in your tests. It takes an explicit effort to identify appropriate triggers and to incorporate them into whatever type of test designs you produce. All of the defect triggers should be included either in a single test design or a set of test designs related to a specific game feature, function, or capability. If you have data from previous projects, see which triggers have been effective at finding defects and include those in your designs, as well as any others you can think of.

The effectiveness of each trigger can be measured in terms of defects per test. You can also think of this as the sensitivity of the game code to each trigger. A large defects/test number relative to other triggers tells you how to find bugs economically and could also hint at an underlying flaw in the game platform design or implementation. If you only have time or resources to run a given number of tests, running the tests for the most effective triggers will yield more defects than running the tests for the trigger that produces the fewest defects per test (usually the Normal trigger). As you continue to create and run more tests for the most effective triggers, you will saturate them and will no longer be able to find new bugs. Repeat this process to establish saturation for all of the triggers.

Combinatorial Design Trigger Examples

Let's go back to the *Halo: Reach* Controller menu combinatorial table, shown in Chapter 8, "Combinatorial Testing" (Table 8.24), to see if any triggers need to be added. Look Sensitivity is tested for its default, minimum, and maximum values. The minimum and maximum values could be considered Stress values because the game is supposed to respond ("process") as slowly or as quickly as it can to the movement of the joystick. The remaining parameters have values that determine whether a capability is either on or off. None of this address a particular configuration or a situation that would apply to Startup, Restart, Exception, or Stress conditions. As a result, the majority of test values represent Normal behavior. For this test to be more effective, incorporate the missing triggers as well as any other possible Stress values.

Start by identifying Configuration resources related to the Controller options. Online players typically use a headset in conjunction with the game controller. This affects where the game audio is routed: to your headset or to your game console's audio output. By design, some audio sources will continue to be routed to your main speakers and others to the headset. The controllers themselves can be wireless or wired. Each controller is sequentially assigned to unique slots on the game console. It is also possible to remove a controller during the process of selecting the options, and subsequently re-connect it in the same position or in a different one. Wireless controllers "disconnect" when going out of range from the console's wireless receiver, when their batteries run out of power, or when the player intentionally removes the batteries. Disconnecting a controller connected to an accessory could have unintended consequences, such as resetting calibration values on a racing wheel. These possibilities suggest new parameters and values to add to the combinatorial table.

The updated table is shown in Table 12.2. Because of the added complexity introduced by the new parameters and values, the Allpairs tool was used to generate this table.

TABLE 12.2 Controller settings combinatorial table with Configuration triggers

Test	Look Sensitivity	Look Inversion	AutoLook Centering	Crouch Behavior	Clench Protection	Remove Controller	Replace Controller	Headset Equipped	Controller Connection
1	1	YES	YES	HOLD	ENABLE	1	1	YES	WIRED
2	3	NO	NO	TOGGLE	DISABLE	1	2	NO	WIRELESS
3	3	NO	YES	HOLD	DISABLE	2	1	YES	WIRELESS
4	1	YES	NO	TOGGLE	ENABLE	2	2	NO	WIRED
5	10	NO	YES	TOGGLE	ENABLE	3	3	YES	WIRED
6	10	YES	NO	HOLD	DISABLE	3	4	NO	WIRELESS
7	1	NO	NO	HOLD	DISABLE	4	3	YES	WIRELESS
8	3	YES	YES	TOGGLE	ENABLE	4	4	NO	WIRED
9	10	NO	NO	TOGGLE	ENABLE	1	1	NO	WIRELESS
10	10	YES	YES	HOLD	DISABLE	2	2	YES	WIRED
11	3	YES	NO	HOLD	ENABLE	3	3	NO	WIRED
12	1	NO	YES	TOGGLE	DISABLE	3	4	YES	WIRELESS
13	10	YES	NO	TOGGLE	DISABLE	4	1	YES	WIRED
14	1	YES	YES	HOLD	DISABLE	1	3	NO	WIRELESS
15	3	NO	NO	HOLD	ENABLE	2	4	YES	WIRED
16	1	NO	YES	HOLD	ENABLE	4	2	NO	WIRELESS
17	10	NO	YES	TOGGLE	ENABLE	2	3	NO	WIRELESS
18	10	YES	NO	HOLD	DISABLE	1	4	YES	WIRED
19	3	YES	YES	TOGGLE	DISABLE	3	1	NO	WIRELESS
20	1	NO	NO	TOGGLE	ENABLE	3	2	YES	WIRED

As an alternative, create a separate table to cover the configuration-related parameters and value pairs. This approach enables you to use the mostly "Normal" table as a reference ("sanity test") and then switch over to the tables for the other triggers once the game passes the sanity tests. The Controller settings configuration table is shown in Table 12.3

TABLE 12.3 Controller Actions configuration table

Test	Remove Controller	Replace Controller	Headset Equipped	Controller Connection
1	1	1	YES	WIRED
2	1	2	NO	WIRELESS
3	1	3	YES	WIRELESS
4	1	4	NO	WIRED
5	2	1	NO	WIRELESS
6	2	2	YES	WIRED
7	2	3	NO	WIRED
8	2	4	YES	WIRELESS
9	3	1	YES	WIRED
10	3	2	NO	WIRELESS
11	3	3	YES	WIRELESS
12	3	4	NO	WIRED
13	4	1	NO	WIRELESS
14	4	2	YES	WIRED
15	4	3	NO	WIRED
16	4	4	YES	WIRELESS

The next step is to seek out Exception trigger opportunities. Because the option values are selected by scrolling, there is no opportunity to enter a "wrong" value. It is perhaps possible to disrupt the selection mechanism itself, however. The A and B buttons are used for accepting the options or going back to the previous screen. Try holding down X, Y, the Left Trigger ("L"), or Right Trigger ("R") during the selection of the test values. Again, one of your options is to add a column for these values, plus the "None" choice, into a single table, as shown in Table 12.4. Although the table has grown again, these 28 cases represent pairwise coverage of 15,360 total possible combinations of these values!

TABLE 12.4 Controller settings with Configuration and Exception triggers

Test	Look Sensitivity	Look Inversionss	AutoLook Centering	Crouch Behavior	Clench Protection	Remove Controller	Replace Controller	Headset Equipped	Controller Connection	Simulta-neous Key
1	1	YES	YES	HOLD	ENABLE	1	1	YES	WIRED	NONE
2	3	NO	NO	TOGGLE	DISABLE	2	2	NO	WIRELESS	NONE
3	10	YES	NO	TOGGLE	ENABLE	1	2	NO	WIRED	X
4	1	NO	YES	HOLD	DISABLE	2	1	YES	WIRELESS	X
5	3	NO	YES	HOLD	ENABLE	3	3	NO	WIRED	Y
6	10	YES	YES	TOGGLE	DISABLE	4	4	YES	WIRELESS	Y
7	1	NO	NO	TOGGLE	ENABLE	3	4	YES	WIRELESS	L
8	3	YES	NO	HOLD	DISABLE	4	3	NO	WIRED	L
9	10	NO	NO	HOLD	DISABLE	1	3	YES	WIRELESS	R
10	3	YES	YES	TOGGLE	ENABLE	2	4	NO	WIRED	R
11	10	YES	NO	TOGGLE	DISABLE	3	1	NO	WIRELESS	Y
12	1	NO	YES	HOLD	ENABLE	4	2	YES	WIRED	L
13	1	YES	YES	TOGGLE	DISABLE	1	3	NO	WIRELESS	NONE
14	10	NO	NO	HOLD	ENABLE	2	4	YES	WIRED	NONE
15	3	NO	YES	TOGGLE	ENABLE	3	1	YES	WIRELESS	X

(Continued)

TABLE 12.4 Controller settings with Configuration and Exception triggers (continued)

Test	Look Sensitivity	Look Inversionss	AutoLook Centering	Crouch Behavior	Clench Protection	Remove Controller	Replace Controller	Headset Equipped	Controller Connection	Simulta-neous Key
16	1	NO	NO	HOLD	DISABLE	1	2	YES	WIRED	y
17	3	YES	NO	TOGGLE	DISABLE	1	1	NO	WIRED	L
18	1	NO	NO	TOGGLE	ENABLE	4	1	NO	WIRELESS	R
19	10	YES	YES	HOLD	DISABLE	3	2	NO	WIRED	X
20	10	YES	NO	HOLD	ENABLE	2	3	YES	WIRED	R
21	10	NO	YES	HOLD	DISABLE	2	4	NO	WIRELESS	L
22	3	NO	NO	TOGGLE	ENABLE	4	3	YES	WIRELESS	X
23	1	YES	NO	TOGGLE	DISABLE	3	2	YES	WIRELESS	R
24	3	YES	YES	HOLD	ENABLE	1	4	YES	WIRELESS	Y
25	10	YES	NO	HOLD	DISABLE	4	4	NO	WIRED	NONE
26	1	NO	YES	TOGGLE	ENABLE	3	3	NO	WIRED	NONE
27	1	YES	NO	TOGGLE	DISABLE	2	4	NO	WIRED	X
28	3	NO	YES	TOGGLE	ENABLE	2	2	NO	WIRELESS	Y

A potential danger in doing this is that most of your test cases will result in an exceptional behavior that might prevent you from observing the effects of the other test values. In Table 12.4, only six tests—1, 2, 13, 14, 25, and 26—avoid a possible input exception. A way around this is to create a separate table to isolate the exception effects, as shown in Table 12.5. The "NONE" value for the Simultaneous Key parameter is not included because it is not an Exception trigger, and it is already implicitly represented in the non-exception table for this feature.

TABLE 12.5 Controller settings table with only Exception triggers added

Test	Look Sensitivity	Look Inversion	AutoLook Centering	Crouch Behavior	Clench Protection	Simultaneous Key
1	1	YES	YES	HOLD	ENABLE	X
2	3	NO	NO	TOGGLE	DISABLE	X
3	1	NO	YES	TOGGLE	ENABLE	Y
4	3	YES	NO	HOLD	DISABLE	Y
5	10	YES	YES	TOGGLE	DISABLE	L
6	10	NO	NO	HOLD	ENABLE	L
7	1	YES	NO	TOGGLE	DISABLE	R
8	3	NO	YES	HOLD	ENABLE	X
9	10	YES	YES	HOLD	DISABLE	X
10	10	NO	NO	TOGGLE	ENABLE	Y
11	1	NO	NO	HOLD	DISABLE	L
12	3	YES	YES	TOGGLE	ENABLE	L
13	10	YES	NO	HOLD	ENABLE	R

The extreme Look Sensitivity values were identified as Stress triggers, but is there any other "stressful" operation that can be done during option selection? For this particular game, both the left analog stick and the D-Pad on the game controller can be used to scroll through the options (vertically) and choices (horizontally). Operating them simultaneously could produce interesting results. Add this to the test table by defining the Scroll Control parameter with the values of LEFT STICK, D-PAD, and BOTH. Follow the same rationale as for the previous triggers when deciding whether to add these parameters and values to a single table for this screen versus creating a separate table for this trigger.

All that is left to do now is to identify Startup and Restart triggers for your Controller settings. These particular settings are tied to individual player profiles. This presents the opportunity to test the settings for a completely new profile versus one that has already been in use. The new profile behavior is your Startup trigger. Add this to the tests as a "Profile" parameter with NEW (Startup) or EXISTING (Normal) choices.

The Controller setting selection process can be restarted in a variety of ways: go back to the previous screen without saving, eject the game disk from the console, or turn the console off. Follow up these operations by going back into the Advanced Controls screen to check for any abnormalities. Because these settings can be stored in either internal or removable memory, another way to do a "restart" is to load information previously saved to external memory back on top of your internally saved modified values. Represent these possibilities in your table with a "Re-Enter" parameter that has a possible value of NONE for the Normal trigger and BACK, EJECT, OFF, and LOAD EXTERNAL for the Restart trigger.

TFD Trigger Examples

TFD triggers are located along the flows. The Ammo TFD template, provided in Appendix E, will be used to illustrate how to incorporate all of the defect triggers into a TFD. It has a few more flows than the TFD you constructed in Chapter 10, but is it "complete" in terms of triggers? Use it in one of the *Unreal Tournament* games and see what you can find.

To begin with, the template includes several Normal trigger flows, such as GetGun and GetAmmo when you have neither (NoGunNoAmmo). The same event can represent different triggers, however, depending on its context with respect to the states it is leaving and entering. For example, GetAmmo when you already have the maximum amount is a case of performing a function when a resource (ammo) is at its maximum. This qualifies as a Stress trigger. Shooting a gun with no ammo falls on the other end of the spectrum where the ammo resource is at a minimum (0). Figure 12.5 shows the Ammo TFD template with these Stress triggers highlighted.

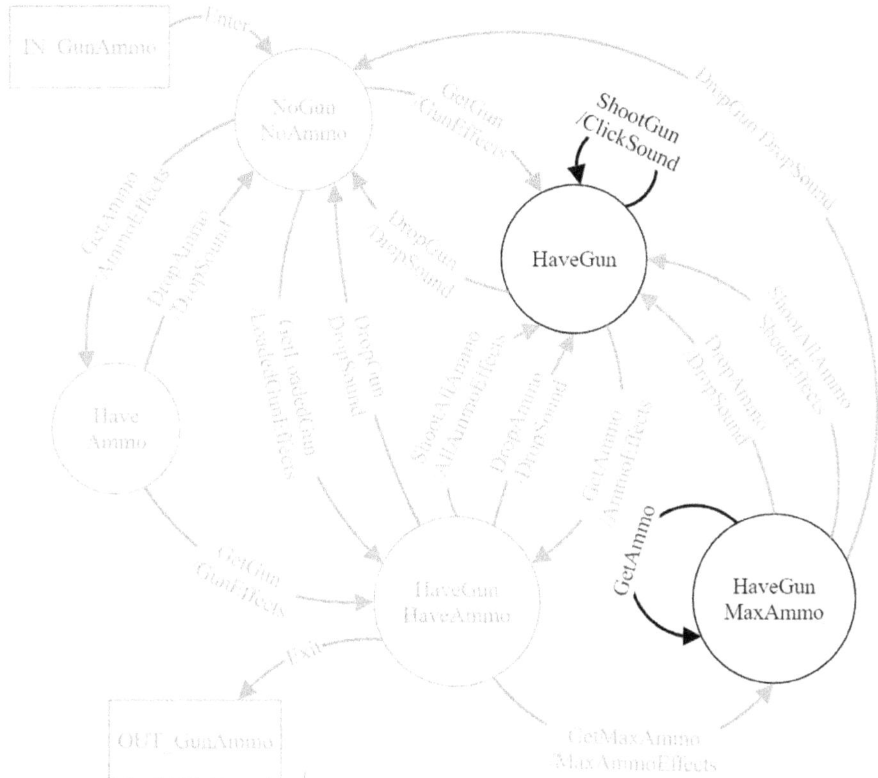

FIGURE 12.5 "Ammo" TFD template with Stress flows highlighted

Now how about adding a Startup trigger? The TFD Enter flow jumps right to the point where the player is active in the match. In reality, there is a "pre-game" period where the player can run around the arena before hitting the "fire" button (usually the left mouse button) to initiate the match. This is relevant to the purpose of the test because a player who runs over weapons or ammo during this time should not accumulate any items as a result.

Represent this startup process on the TFD by performing "mitosis" on the "NoGunNoAmmo" state: That is, split it into two connected states. One state retains the original name and connections (except for the "Enter" flow) while the other captures the dry run and countdown behavior. Figure 12.6 shows the process of splitting this portion of the TFD.

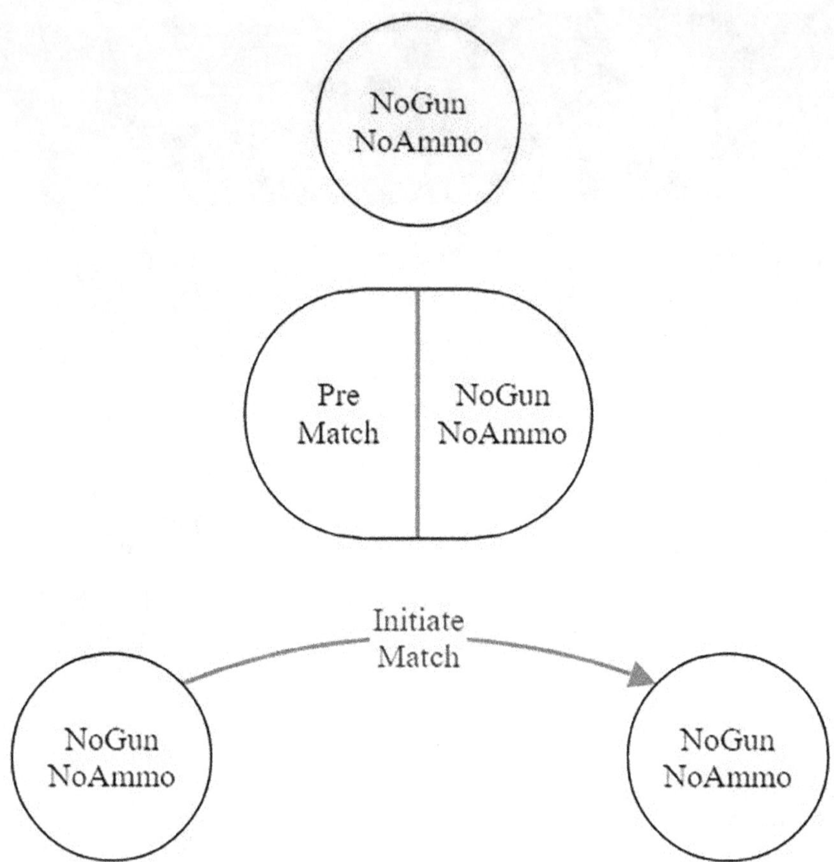

FIGURE 12.6 "NoGunNoAmmo" state "mitosis"

The new "PreMatch" state can be introduced to the TFD. Start by disconnecting the "Enter" flow from "NoGunNoAmmo" and attaching it to "PreMatch." Then add flows to attempt "GetAmmo" and "GetGun" during the "PreMatch" period. These flows are Startup triggers, as shown in Figure 12.7.

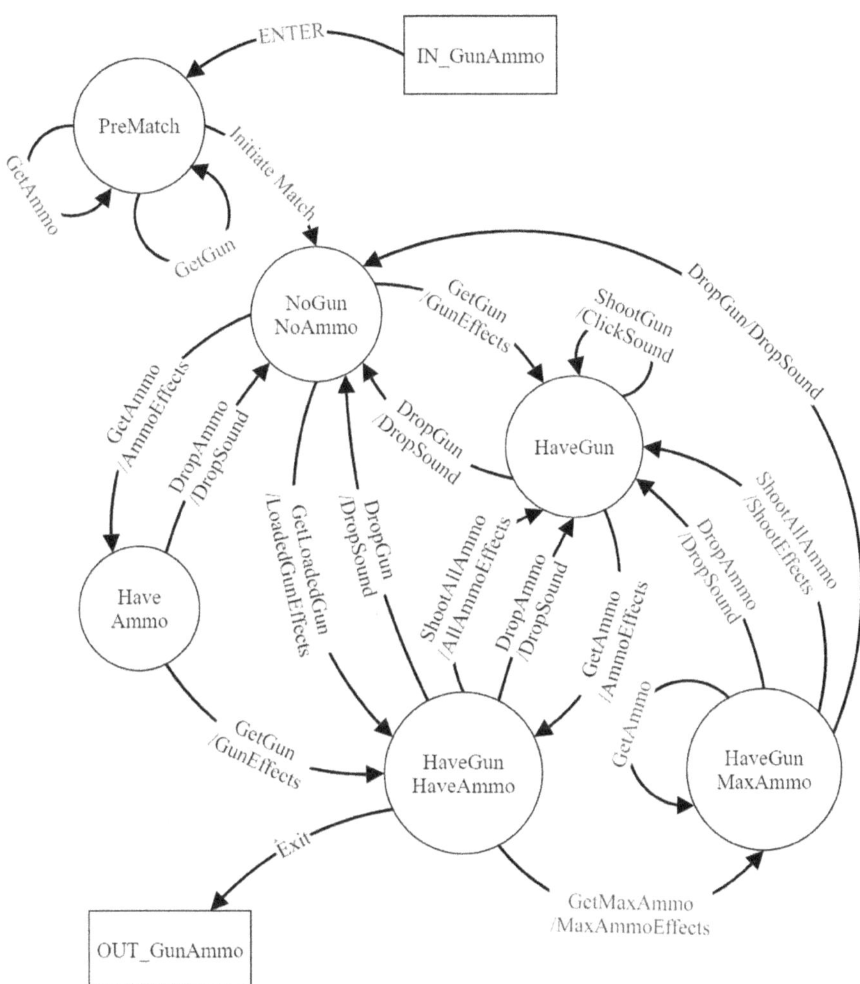

FIGURE 12.7 "PreMatch" state and Startup trigger flows added to "Ammo" TFD

Next, add the Restart trigger to the diagram. It is possible to change your status to "Spectator" in the middle of a match, and then join back in as a participant. Spectator mode takes your character out of the game and lets you follow players in the game while you control the camera angle. Any guns or ammo picked up prior to entering Spectator mode should be lost when you join the same match that is still in progress. Rejoining the game from Spectator mode is done instantly without the countdown timer that you get when you start the match for the very first time. Suspending and rejoining can be done at any time during the match after the initial countdown timer has expired. Add a "SpectateAndJoin" flow from each of the in-game states on the TFD and tie it back to "NoGunNoAmmo." Do not forget the loop flow from "NoGunNoAmmo" back to itself. A TFD with these updates is shown in Figure 12.8.

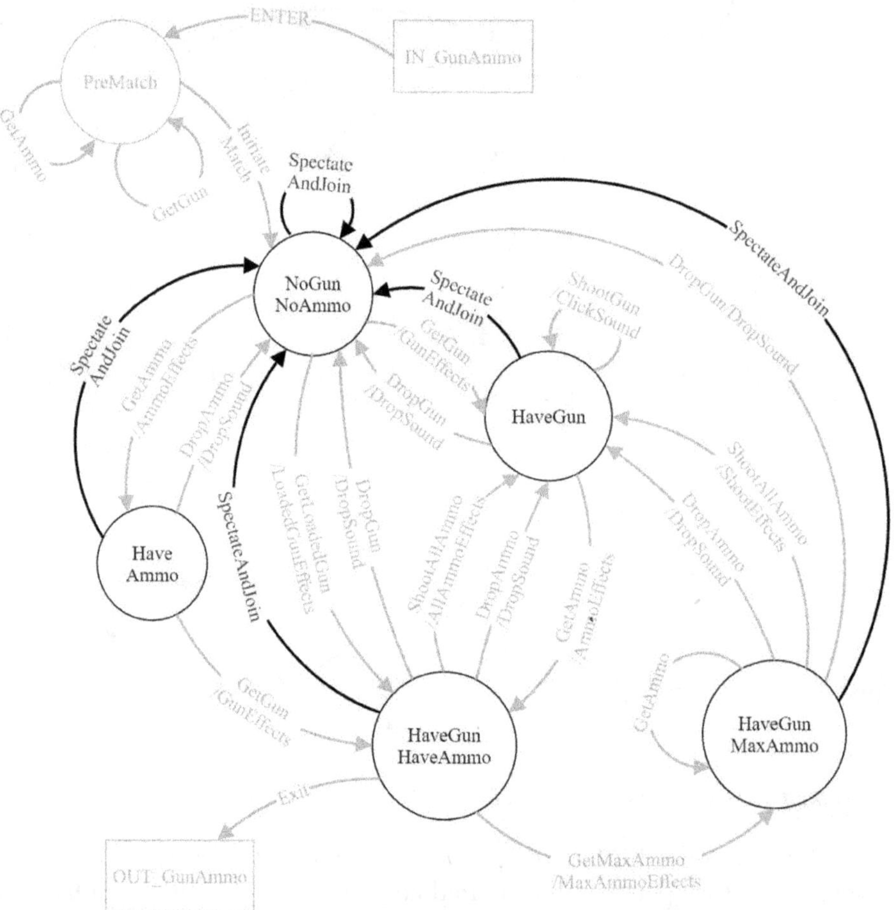

FIGURE 12.8 "Ammo" TFD with "SpectateAndJoin" Restart flows highlighted

Note that more "pressure" is being put on the "NoGunNoAmmo" state with all of the flows entering and exiting. It is like a busy intersection; they tend to be much more dangerous than the ones that are not so busy.

This reflects the importance of this state to the well-being of the feature and its potential sensitivity to changes.

There are still a few more triggers to consider for the TFD. For the Exception trigger, there is an operation available to use a weapon's alternate-fire mode. Typically, the left mouse button is used for normal firing and the right mouse button for alternate firing. Some weapons, such as the Grenade Launcher, do not have an alternate firing mode. They should not fire nor decrement their ammo count when the user attempts alternate firing. This is something you can use as an Exception trigger. Because this "UnsupportedAltFire" operation will not change the ammo status of the weapon, add it as a loop on the TFD states where there is both a gun and ammo. Your result should resemble the diagram in Figure 12.9.

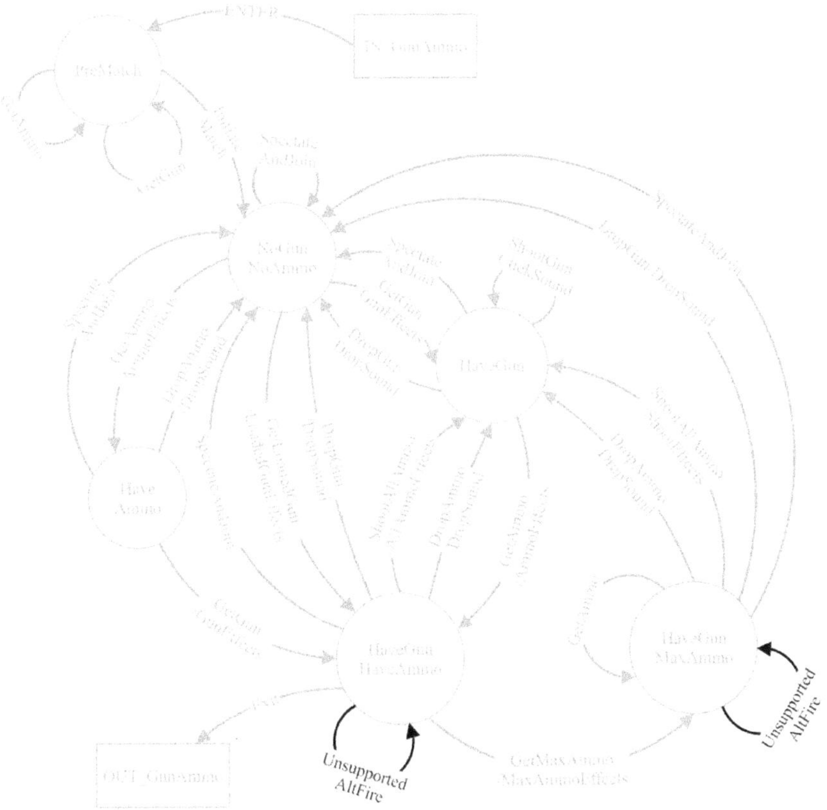

FIGURE 12.9 "Ammo" TFD with "AltFire" Exception flows highlighted

Finally, the Configuration must be included. One of the weapon settings in the game allows the player to select an older style rendering of certain weapons. Although this is appealing to players who owned earlier titles in this series, it also creates an additional test responsibility. You can check that changing the weapon rendering while the game is in progress does not affect the amount of ammo loaded in that weapon nor produce unwanted artifacts, such as unexpected audio or "shadow" (duplicate) weapons. Add a "ToggleWeaponSkin" flow at all of the states where the player has the weapon. Because this should not affect the ammo, these flows will loop back into the states from which they originated. Figure 12.10 shows the TFD with these Configuration flows.

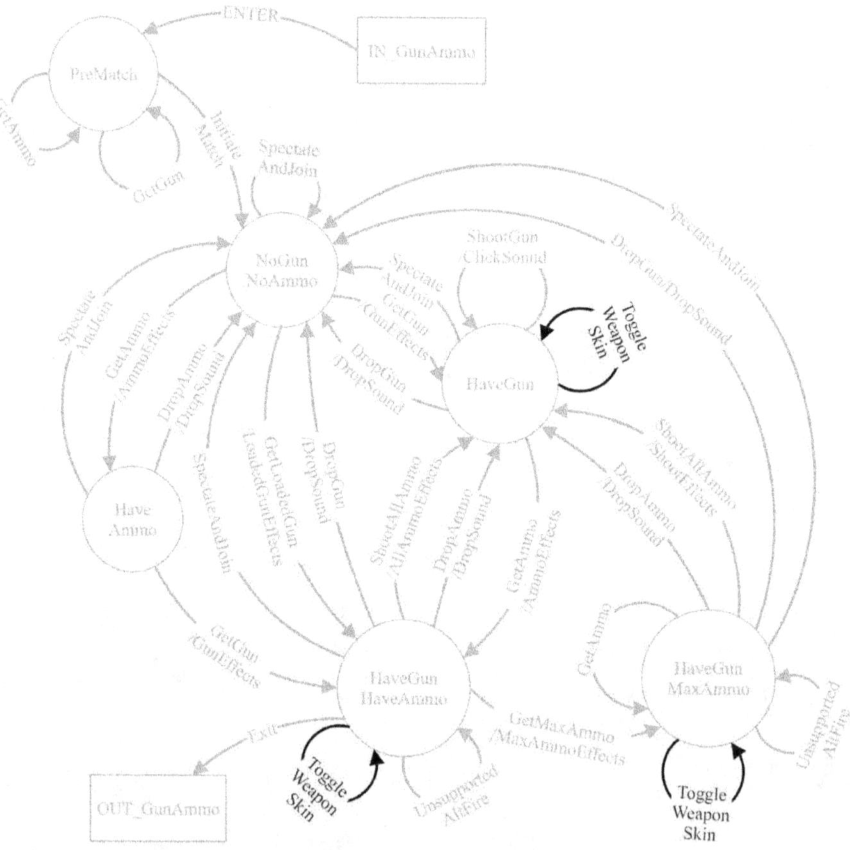

FIGURE 12.10 "Ammo" TFD with weapon skin Configuration flows

Now the TFD is getting really crowded! Just remember that the same option that was presented for combinatorial tests is applicable to TFDs, test tree designs, or any other formal or informal means you use to come up with test cases. You can incorporate the triggers into a single design or create companion test designs that work together as a suite of tests to provide the trigger coverage you need.

You might also find it useful to document the intended triggers for each test element. One easy way is to provide a letter code in parentheses after the event name on each TFD flow, parameter value for combinatorial tests, or branch node for test tree designs. You can count the number of times each letter appears to see how many times each trigger is being used. It also helps you classify defects you find when running the tests. Just be aware that this carries a maintenance burden to reevaluate the trigger designation whenever you move or add new test elements to the design.

What a difference the extra triggers make in the test design! Is it more work? Yes. It is also better. You have improved the capability of this test to find defects, and you will have more confidence in your game when it passes tests that use all of the triggers. Defect triggers were not created with any one particular test methodology in mind. They are effective whether you are testing at the beginning or the end of the project and whether you have meticulous test designs or you are just typing in test cases as you go along. If you choose not to use them, you are adding to the risk of important defects escaping into your shipping game.

EXERCISES

Having come this far in the book, you should be well equipped to offer your own suggestions and to contribute to your team's test strategies. The following exercises are designed to give you some practice at that.

1. Which is your favorite defect trigger? Why? Which one would be the most difficult for you to include in your tests, both in terms of test execution and test design?
2. Earlier in this chapter, we mentioned that both the D-Pad and Analog joystick on the game controller could be used to make the *Halo: Reach* option selections. Describe how you would incorporate these choices into your test suite. Do you prefer adding them to a large single table

for the feature or creating a separate smaller table focused on the option selection means? Be specific about which factors would cause you to change your answer.

3. It could be interesting to start an *Unreal Tournament* match while standing on one of the gun or ammo items. The game automatically snaps you back to the original starting point after a 3-second countdown before the action starts. Describe how you would update the "Ammo" TFD to include this possibility, including what effects you would check for and why.

4. Again, for the "Ammo" TFD, describe how you would add or change flows to represent someone playing from a PC who can fire her gun using either a joystick or the left mouse button. Treat this as a case where both the mouse and joystick are connected during the game. Indicate which triggers are represented by this possibility.

5. Make a list or outline of how you would include each trigger in your testing of a hypothetical or actual *Texas Hold 'Em* video game. Do not stop at one example—list at least three values, situations, or cases for each of the non-Normal triggers. Remember to include tests of the betting rules, not just the mechanics and winning conditions for the hand. If you are not familiar with the rules of this card game, do a search online and read a few descriptions before you build your trigger lists.

REFERENCE

Nexus Mods. 2008. "Official Oblivion 12416 Patch." Last modified June 26, 2008. *https://www.nexusmods.com/oblivion/mods/11364*.

CHAPTER 13

Regression Testing and Test Reuse

REGRESSION TESTING

Regression Testing is a strategy for deciding which tests to run against each version of the game. This applies to code that is under development as well as to bug-fix releases. It gets its name from the need to determine if any of the code has "regressed" (gone backward in progress) due to changes introduced in the latest build. A good strategy will minimize the number of tests you run and will still be able to help you catch newly introduced and remaining errors.

Chapter 5, "The Phases of Game Quality Assurance," describes the important role regression testing plays in distributing a good release. Once testing is under way, you need to be able to react in real time as the game code or assets become updated in response to bugs or change requests. You also need to be able to adjust your tests to cover new changes in the code or specs.

The A-B-C Method

Regression testing needs to do more than re-run tests that have previously failed. The rationale for this is based on the phenomenon of "*software rot*," which can be categorized by two types:

Dormant rot refers to software that is not used on a consistent basis, making it prone to become useless as the rest of the application evolves.

Active rot occurs when constant modifications and bug fixes gradually affect the integrity of the original feature or code base.

Every time you are given new code to test, you are getting a combination of code that has not been modified, code that has intentionally changed, and code that perhaps was unintentionally affected by the changes.

One approach for combating software rot is to break your tests in thirds, executing a new third every time you get new code. Rather than breaking up your entire test suite into top, middle, and bottom thirds, it is better to divide each major function or feature into thirds. This helps re-establish that each feature is working correctly in every build, demonstrates that the code has not significantly decayed, and keeps your test results up to date.

Here is an example of how you could distribute your tests across a mobile card battle game where you have one screen for purchasing cards, another for assembling your deck, and another for battling with an AI opponent and determining a winner.

In this imaginary test suite, there are 40 combinatorial test cases for purchasing cards, two TFDs with a total of 20 paths for assembling your deck, and one TFD with 6 paths and 15 tests written manually to test the card battle process. Table 13.1 shows how these can be distributed into three test sets called A, B, and C.

TABLE 13.1 A-B-C Distribution of card battle tests

Card Battle Feature	Test Type	"A" Tests	"B" Tests	"C" Tests
Purchase Cards	Combinatorial	13	13	14
Deck Forming	TFD	7	7	6
Resolve Battle	TFD + Manual	7	7	7

To distribute the tests even more efficiently, break them down by each design. If the three "Purchase Cards" combinatorial designs generate 12, 12, and 16 test cases respectively, then the "A" tests should use four tests from the first combinatorial table, four from the second, and five from the third. If the two "Deck Forming" TFD designs have 11 and 9 paths respectively, then use four from the first TFD and three from the second one. Follow suit for the "Resolve Battle" tests. Then repeat the process for the "B" and "C" cycles so that all of the test cases are scheduled to be run across all three cycles. Table 13.2 shows a detailed breakdown and A-B-C distribution of the "Purchase Cards" combinatorial test cases generated by separate designs for selecting a pack to purchase, paying for the cards, and updating the card inventory.

TABLE 13.2 Detailed distribution of "Purchase Cards" combinatorial test cases

Test Design	A Cycle	B Cycle	C Cycle
SelectPack	Test1		
	Test2		
	Test3		
	Test4		
		Test5	
		Test6	
		Test7	
		Test8	
			Test9
			Test10
			Test11
			Test12
PayForCards	Test1		
	Test2		
	Test3		
	Test4		
		Test5	
		Test6	
		Test7	
		Test8	
			Test9
			Test10

(Continued)

TABLE 13.2 Detailed distribution of "Purchase Cards" combinatorial test cases (continued)

Test Design	A Cycle	B Cycle	C Cycle
			Test11
			Test12
UpdateInventory	Test1		
	Test2		
	Test3		
	Test4		
	Test5		
		Test6	
		Test7	
		Test8	
		Test9	
		Test10	
			Test11
			Test12
			Test13
			Test14
			Test15
			Test16

A good procedure to follow whenever you get a release with bug fixes is to run the tests that previously failed plus any new tests you created for that bug, regardless of which cycle you are on. Following that, run the tests you identified for the current cycle to maintain your confidence in the quality of the remaining features.

Defect Modeling

Besides deciding which tests to run, regression testing can also involve the modification of existing tests or the creation of new tests. When you do not have specific tests for new issues or tweaks that pop up in Alpha, Beta, post-release, or patches, you can provide targeted test designs that model specific bugs or changes. Like the tests you created earlier in the project, the new tests should be run in cycles to establish that the changes continue to work as intended. As an example, we will create a test to model a defect that was fixed in the Title Update 2 patch for *Gears of War 2* (Fandom.com n.d.):

```
An issue that could cause players' Look Sensitivity to be
changed to their Zoom sensitivity while zooming in, zooming
out, and firing.
```

Gears of War 2 gives the player the ability to configure three independent weapon-wielding sensitivity parameters. The *Look Sensitivity* affects how quickly your player can turn back and forth when looking around the environment with his weapon at his side. *Target Sensitivity* determines how fast you can swivel your weapon when it is raised and you are looking through the gun sight for targets to shoot. The *Zoom Sensitivity* value determines how quickly your player can turn when your weapon is raised and zoomed in to magnify your target. For example, when wielding a sniper rifle, you might want to be able to look around quickly to find something to shoot at, have some more control when you are picking out a distant target, and slowly guide the crosshairs to get a precise shot when you zoom in with the scope. In that case, you would have Look Sensitivity = High, Target Sensitivity = Medium, and Zoom Sensitivity = Low. When the defect changes the Look Sensitivity to the Zoom Sensitivity, that will slow down your ability to scan and notice incoming enemies when you are not aiming at anything, contrary to your original setup.

To get this TFD started, draw bubbles and connecting lines that exactly match the different states of the game that are described in the bug report. The report explicitly mentions three events: zooming in, zooming out, and firing. It is logical to model this with a "zoomed in" state and a "zoomed out" state. Events like "FireWeapon" and "ZoomIn" provide transitions and loops. Figure 13.1 shows a first cut at the sensitivity regression scenario.

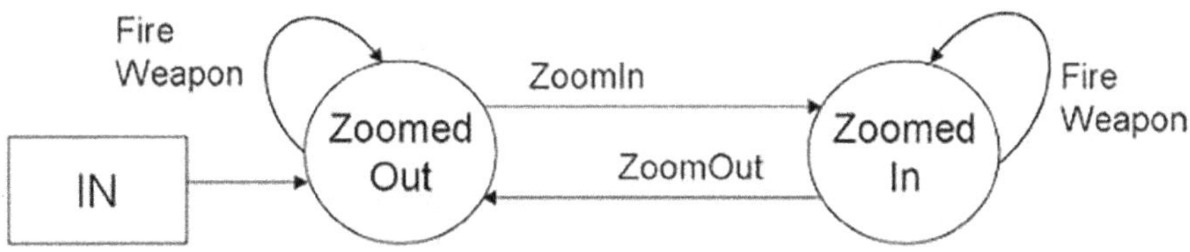

FIGURE 13.1 First stage of "Look Sensitivity" bug-fix verification TFD

This is a good start, but this design does not have any states or flows related to the use of the "Look Sensitivity" parameter, which is the part that was broken prior to the patch. Because "Look Sensitivity" affects the player's turning speed when their weapon is lowered, this test needs a "NotAiming" state. That will serve as a place to start the failure verification scenario and a place to return to in order to check that "Look Sensitivity" behaves according to the player's setting once he lowers his weapon. The "NotAiming" state is also a good place to connect the "OUT" box because it will force the "Look Sensitivity" to be checked at the end of each test.

Finally, this TFD needs to take into account what should be verified when arriving at each zoomed state. The tests generated from this design must check that the appropriate sensitivity setting is applied to each zoomed state. That checking can be done each time the test arrives at the "ZoomedIn" or "ZoomedOut" by performing a "look" so the data dictionary definitions for those states must instruct the tester to look and check that the appropriate sensitivity is used—the "Zoom Sensitivity" value for the "ZoomedIn" state, the "Target Sensitivity" for the "ZoomedOut" state, and the "Look Sensitivity" for the "NotAiming" state.

Figure 13.2 shows the full verification TFD with actions, flow numbers, and the "OUT" state added to the diagram.

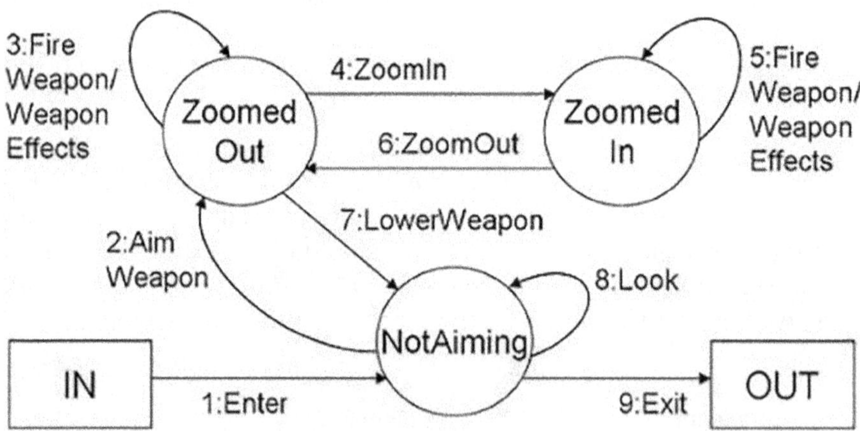

FIGURE 13.2 Complete "Look Sensitivity" bug-fix verification TFD

Once the diagram and checks are in place, you need to develop a set of paths to test. The situation here is slightly different from when you create a fresh new TFD for something you have not tested yet. In this case, you must test the specific path that matches the failure scenario described in the bug report. If there was more than one way to cause the failure, each of those paths must be defined for this test. Once that is done, define additional paths to ensure each flow is tested at least once. For the current design, the defect verification paths should include the zoom in/zoom out/fire sequence, which corresponds to flows 4, 6, and 3, plus doing the "Look" after lowering your weapon to make sure "Look Sensitivity" has not changed. This makes the basic bug verification path 1, 2, 4, 6, 3, 7, 8. This path should also function as the baseline path for this TFD.

Beyond the baseline, you need to have at least one path that includes flow 5, but also consider what other paths will do a thorough job of ensuring that the original defect is not still lurking in the code. Be sure to define one or more paths with loops, cycle through the baseline and a take a long path or two. Some paths that would be appropriate for the current example are as follows:

i. *Loops*— Fire your weapon multiple times when zoomed in, then zoom out, and lower the weapon to check "LookSensitivity."

j. Follow the baseline to the "ZoomedIn" state, then "ZoomIn" and "ZoomOut" multiple times before firing from the "ZoomedOut" state.

k. After entering, look multiple times from the "NotAiming" state before proceeding with the rest of the baseline.

Baseline Cycling—Follow the baseline and return to the "NotAiming" state, but instead of exiting from there, go back through the baseline a second time.

Long Path—Go through each flow three times in different sequences.

The benefit of modeling the defect and doing a new design instead of running an existing test that mostly covers the defect scenario, is that you will do a better job of revealing manifestations of related defects that could exist but were not found during the initial testing or new defects that were introduced by the code changes that were made to fix the bug. The purpose is to create a safety net of tests that will increase your confidence that the original issues are fixed and no new issues were introduced by the fixes.

Time Keeps on Ticking

Some games seem as though they have been played almost forever. Successful sports titles in particular can evolve and grow over a course of many years by updating rosters, uniforms, schedules, stadiums, and so on.

You can play one version of the game over many seasons or purchasing the newer editions of the game. When the same part of a game is exercised over and over again, unintended side effects could appear. Rather than clearing memory and deleting saved files each time you re-run a test, have a machine or drive that keeps saved files so you can accumulate information in order to age the game to the point where the game play starts to break down.

Take a moment to stop and think about detrimental situations that could result from playing or testing a game over a long period of time.

Have you thought about it? Some plausible examples are:

- The weapon shop runs out of inventory.
- The game will not allow you to plant any more trees.
- Your player accumulates money or points until there are no more digits left to represent an increase in the amount.
- All vehicles are damaged so badly that they cannot be used, preventing access to the next area or zone.
- Your stats are maxed out so buffs and bonuses have no effect.

A real-life example of an aging problem shows up in *FIFA 11* after you take your Virtual Pro through many seasons. New players are generated to simulate what happens in real life when the original players do not have their contracts renewed or they retire from the game. The new players receive newly generated names and are incorporated into each of the teams over time. One impact of the newly minted players is that there are no audio assets for the pronunciation of their names, so the in-game announcers continue to use the audio for the original player's name at that roster spot. An additional side effect occurs where some new players are assigned empty names. This manifests itself in many places where the player's name is blank on various game screens and reports.

Figure 13.3 shows an exhibition match lineup screen where the left center midfielder (LCM) for the Bohemians club has a blank name. Wherever the game UI would normally make reference to the player's name, such as when the player scores a goal or receives a yellow card, you see only a blank space or just the player's jersey number. Looking across all of the teams in the game reveals that many other teams exhibit the same problem and some have more than one player without a name.

The lesson here is that if you end your testing after only one season, or never play through the game multiple times, you can miss faults that occur and accumulate over time.

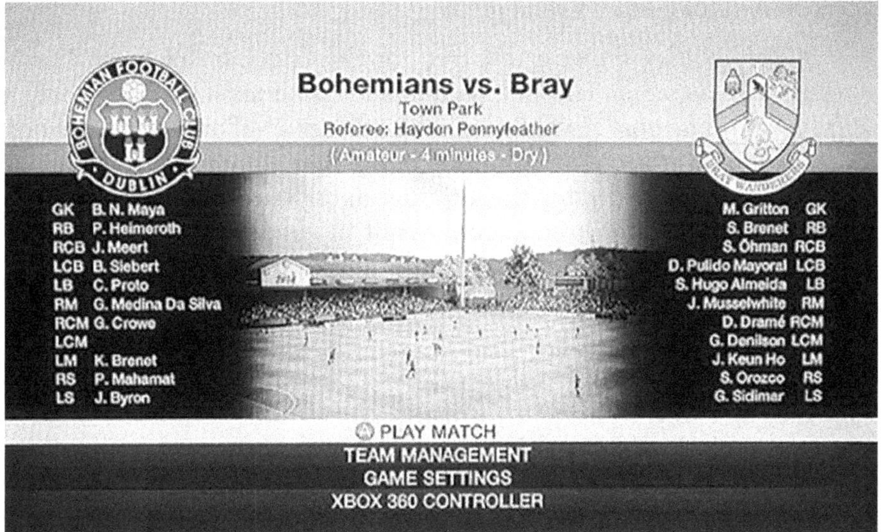

FIGURE 13.3 *FIFA 11* match lineup with a nameless player

Expanding Possibilities

Regression testing also applies to checking original game features in the presence of new additions such as expansion packs or items added to an online store. Integrating them with existing tests might not always be possible.

The *Gears of War 2* "All Fronts" expansion pack added 19 maps, a new single-player chapter, and 13 new achievements. Some of the achievements have requirements that combine original *Gears 2* features or achievements with the new content. For example, the "Be Careful What You Wish For" achievement requires you to have reached level 8 in multiplayer and completed waves 1 through 10 on the Highway map in Horde. A player who had previously reached level 8 or higher should be able to earn the achievement by completing the 10 waves. Players who bought the game and the expansion together, however, would have to satisfy both requirements before receiving the achievement. It is possible for the player to reach the required level before completing the 10th Horde wave, and it is also possible for the player to complete the 10th wave before reaching level 8. A player who is already at level 8 before installing the expansion pack only has to worry about completing the 10th wave on the new map. This situation turns out to be a good time to use a test tree to represent the various ways to fulfill the conditions of the achievement.

Figure 13.4 shows a test tree for the "Be Careful What You Wish For" achievement. The tree accounts for what level the player is at with respect to the goal (level 8), when the target level or target wave criteria is met; it also accounts for both being met simultaneously and which criteria is reached after the first one is met. Reaching any of the terminal nodes in the tree should result in successful completion of the achievement.

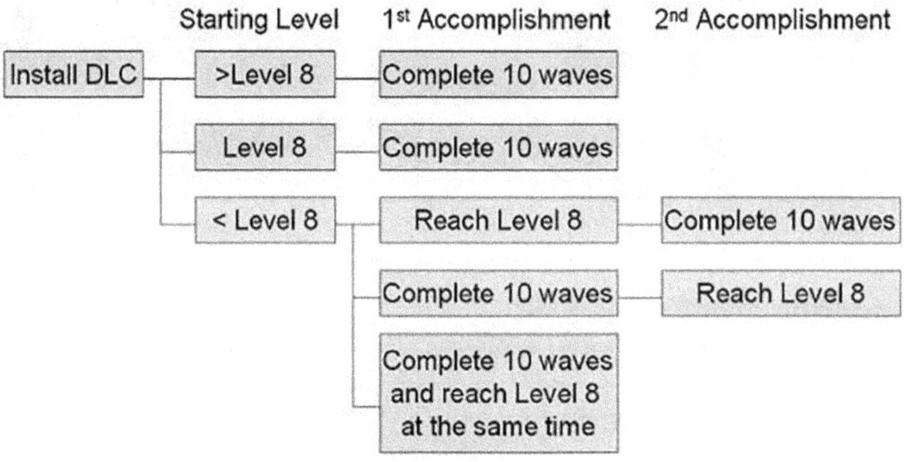

FIGURE 13.4 *Gears of War 2* Highway map achievement tree

TEST REUSE

Test reuse is a strategy to create tests that are designed and structured to expand and adapt to the evolution of the game. The effort you put into the development of a test design or script can be used over and over again for more than one feature, more than one version, more than one game or all of the above. To be successful, you need to think about reuse from the time you begin designing your tests.

TFD Design Patterns

As you gain experience testing games, you will be able to recognize recurring situations within each game, as well as common situations that appear across multiple genres and titles. This provides an opportunity to optimize the way you produce new test designs. Many of these situations can be represented by two or three major game states, with a few transitions between each state. Test Flow Diagrams (TFDs) are a good vehicle for turning out tests based on these patterns. Each new test can be created by changing the state names and flows without having to rethink the structure of the diagram each time. Figure 13.5 shows the framework for a two-state pattern.

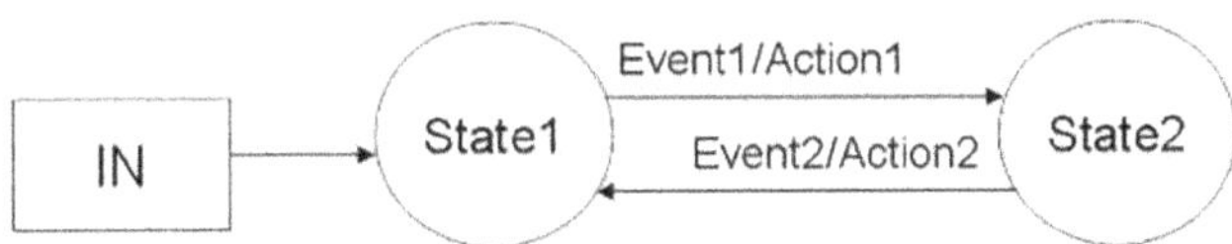

FIGURE 13.5 Two-state TFD design pattern

Here is how to use the pattern for testing weapon swapping in a first-person shooter. "State1" becomes the state where you are wielding the default weapon, so you can call that state "Weapon1." "State2" is where you are using an alternate weapon instead of the default. You get from "State1" to "State2" by swapping weapons, and likewise to get from "State2" back to "State1." This pattern also can be used for swapping weapons from your inventory or having to drop your weapon to pick up a new one. Figure 13.6 shows the weapon swap scenario implemented using the two-state pattern.

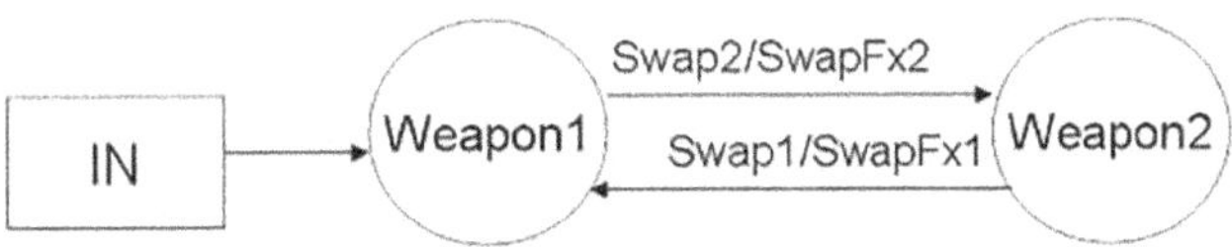

FIGURE 13.6 Weapon swap using the two-state design pattern

To reuse this pattern for the same game running on different platforms, you need to account for platform-specific events. For example, game controls can be very different for a version that is published for consoles versus the same game running on PCs or mobile devices. Swapping weapons might be accomplished by a single controller button hitting a keypad number or tapping on the screen. The differences can be reflected in the name you give to each event, or you can use a generic name for each event and have the flexibility to define whether it is the same button used for each or a different type of control altogether. A simple table can provide the separate event and action definitions for each supported platform without requiring you to make any new diagrams. Table 13.3 provides some example definitions for swapping weapon 1 ("Swap1") on those various platforms:

TABLE 13.3 "Swap1" event definitions for various platforms

Platform	Swap1 event definition
PC	Press the "1" key on the numeric keypad
Console	Press the "X" button on the controller
Mobile	Tap the "1" icon in the lower let corner

Similarly, the actions resulting from each swap could vary due to different animations and sound effects designed for each weapon type or platform. This does not just apply to weapons: you can take the same approach if you are swapping kittens, skee balls, or golf clubs. Use the patterns to get started quickly and turn them into well-designed TFDs by adding related flows and states that are particular to your game and the intended purpose of the test. Additional techniques described in previous chapters, such as expert paths and flow usage profiles, will help you round out your design. Complete the TFD by adding the "OUT" state, numbering the flows, and providing percentages for each flow if you are doing usage-based testing.

Once you generate your paths, your new test is ready to go. Figure 13.7 illustrates some example scenarios that fit into the two-state pattern. This is not an exhaustive list but is meant to encourage you to recognize situations where you can use these patterns to test your game features more efficiently. Note that some scenarios begin with a negative state (e.g., "NotPoisoned") and others with a positive one ("HasBall"). Where you start the test depends entirely on the initial state of the situation you want to model, either by your choice as a test designer or to reflect the natural progression of the game. For example, in the case of the Wolfman path, the game story begins with the hero in human form and the moon not yet full.

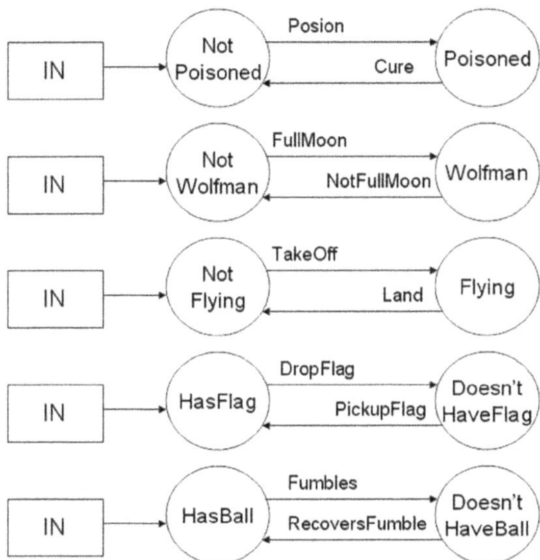

FIGURE 13.7 Example two-state scenario starter diagrams

There are also game situations that can be tested using a three-state pattern. The principle is the same as for the two-state patterns. An extra twist to the three-state template is that one of the states has a transition in only one direction and another always has two flows going back to the starting state. This might not apply one hundred percent of the time, but start with the basic pattern and force yourself to discover what information can be put on each of the flows. After establishing the basic pattern, provide the additions or subtractions that will make your test complete and correct, while keeping it relatively simple. Figure 13.8 shows the template for the three-state TFD pattern.

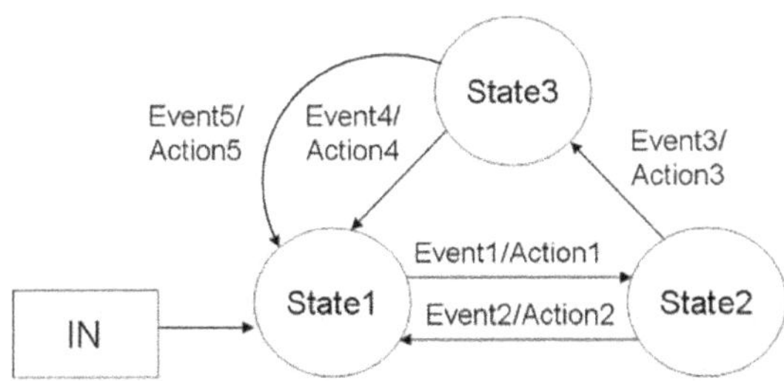

FIGURE 13.8 Three-state TFD design pattern

As with the two-state patterns, you can have different event, action, or state definitions for the same game running on different platforms, but you do not have to change the diagram to accommodate that. Figure 13.9 shows the three-state pattern applied to a few generic game scenarios.

None of the two-state or three-state pattern lists are exhaustive, and the patterns themselves do not represent every game scenario you will encounter. As you progress through your career as a tester, pay attention to the patterns that emerge from your own test designs and add them to your pattern collection. Reuse them and evolve them to get more and more mileage for your effort. At the same time, resist the temptation to be satisfied with a pattern-based test that might not be the best test for the situation at hand. Treat the pattern as a starting point rather than an endpoint. It is not meant to constrain you, but to get you going quickly with the basic patterns so you can direct your energy at including all the special parts that make the test uniquely yours.

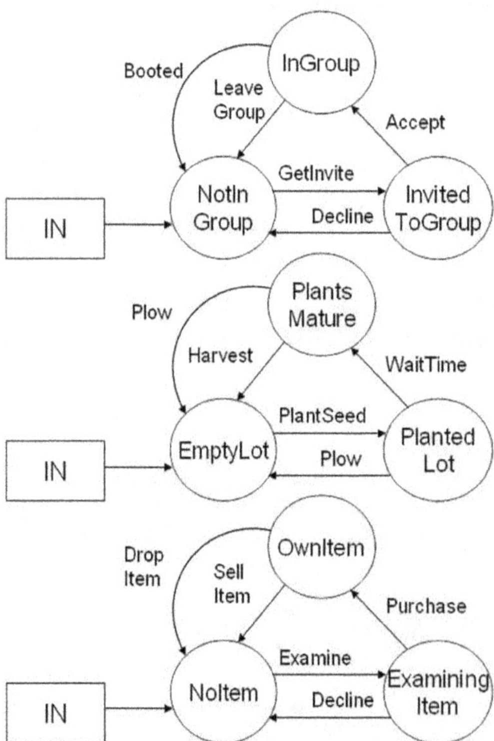

FIGURE 13.9 Example three-state scenario starter diagrams

Looking Back and Forth

Looking back at the TFD created for the *Gears of War 2* sensitivity bug, doing a thorough and efficient job of testing means reusing that one design to test each of the five weapons in the game that have zoom capabilities: Hammerburst Assault Rifle, Longshot Sniper Rifle, Boltok Pistol, Gorgon Pistol, and Snub Pistol. An efficient way to handle this without making edits, creating additional diagrams, or adding paths is to repeat the tests for each zoomable weapon. When the test is reused in this way, each variant needs to show up as a separate item in your test inventory so that you can track results and include each of them in your A-B-C execution cycles. It is also best in this case to mix weapons and paths into each of the cycles so you do not overlook a fault that occurs for an individual weapon or path. Table 13.4 shows how this might look if you have defined three paths for your "Look Sensitivity" TFD.

TABLE 13.4 "Look Sensitivity" test scheduling for zoomable weapons

Test Design: Look Sensitivity		
A Cycle	**B Cycle**	**C Cycle**
Path 1 - Hammerburst		
Path 2 - Longshot		
Path 3 - Boltok		
Path 1 - Gorgon		
Path 2 - Snub		
	Path 3 - Hammerburst	
	Path 1 - Longshot	
	Path 2 - Boltok	
	Path 3 - Gorgon	
	Path 1 - Snub	
		Path 2 - Hammerburst
		Path 3 - Longshot
		Path 1 - Boltok
		Path 2 - Gorgon
		Path 3 - Snub

Earlier in this chapter, a test tree was used to provide a design for testing a particular achievement added to *Gears of War 2* via an expansion pack. Even though this happened after your main test development campaign, you can still benefit from taking a reusable approach. The expansion pack has a total of seven new achievements with similar requirements, varying by what level the player achieves, how many waves must be cleared, and which map must be used. Instead of creating seven similar test trees, produce a single generic version that can be understood and run by testers according to the different data requirements for each achievement. Figure 13.10 shows a generic test tree to accommodate existing and future achievements that are structured the same way.

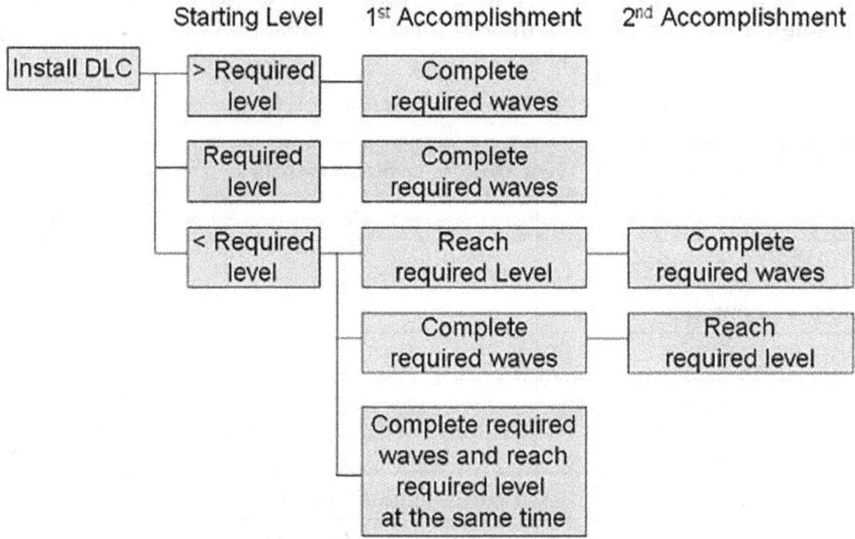

FIGURE 13.10 Generic wave completion achievement tree

Combinatorial Expansion

Games will grow in complexity as they evolve and get updated. Redoing your tests from scratch is not a productive strategy. Because combinatorial tests can absorb new parameters and values with minimum growth of your test inventory, they are a great way to evolve your tests to keep up with new or updated features.

For many years, the *FIFA* franchise from Electronic Arts provided more features to each year's game to keep up with soccer fans' growing expectations. *FIFA 2007* for Xbox 360 provided the following choices:

- Game Settings—Visual screen:
- Time/Score Display: OFF, ON
- Camera: Dynamic, Dynamic V2, Tele, End-to-End
- Radar: 2D, 3D, OFF
- Match Intro: ON, OFF

These values can be reduced to 12 tests using a combinatorial design generated by the Allpairs tool, as shown in Table 13.5.

TABLE 13.5 *FIFA 2007* Visual Settings combinatorial table

	Camera	**Radar**	**TS Display**	**Match Intro**
1	Dynamic	2D	OFF	OFF
2	Dynamic	3D	ON	ON
3	DynamicV2	2D	ON	OFF
4	DynamicV2	3D	OFF	ON
5	Tele	OFF	OFF	OFF
6	Tele	2D	ON	ON
7	EndToEnd	OFF	ON	ON
8	EndToEnd	3D	OFF	OFF
9	Dynamic	OFF	OFF	ON
10	DynamicV2	OFF	ON	OFF
11	Tele	3D	ON	OFF
12	EndToEnd	2D	OFF	ON

For the 2008 edition, the Match Intro setting was removed and two new settings were added: HUD and Indicator. HUD choices are "Player Name Bar" and "Indicator." "Indicator" choices are "Player Name" and "Player Number." The Camera choices were moved to the "Game Settings–Camera" screen and a "Pro" camera choice was added. Separate Camera setting choices can be made for different game modes: Single Player, Multiplayer, Be a Pro, and Online Team Play. For this example, we will treat them as a single setting. When it is time to test, go through the table for each of the different modes, taking advantage of yet another way to reuse this test design.

The 2008 version of the table can be constructed from scratch, but it is easier to just make the changes to the 2007 file you ran through Allpairs and regenerate a new table. Listing 13.1 shows the *FIFA 2008* data for the Allpairs input file and Table 14.6 shows the new visual settings combinatorial table generated for FIFA 2008. One hundred and twenty combinations are now reduced to 16 test cases.

LISTING 13.1 Allpairs File for FIFA 2003 Visual Settings Changes

Camera	Radar	TS-Display	HUD	Indicator
Dynamic	2D	OFF	NameBar	Name
DynamicV2	3D	ON	Indicator	Number
Tele	OFF			
EndToEnd				
Pro				

TABLE 13.6 *FIFA 2008* Visual Settings combinatorial table

	Camera	Radar	TS Display	HUD	Indicator
1	Dynamic	2D	OFF	NameBar	Name
2	Dynamic	3D	ON	Indicator	Number
3	DynamicV2	2D	ON	NameBar	Number
4	DynamicV2	3D	OFF	Indicator	Name
5	Tele	OFF	OFF	NameBar	Number
6	Tele	OFF	ON	Indicator	Name
7	EndToEnd	2D	OFF	Indicator	Number
8	EndToEnd	3D	ON	NameBar	Name
9	Pro	2D	ON	Indicator	Name
10	Pro	3D	OFF	NameBar	Number
11	Dynamic	OFF	OFF	Indicator	Name

(Continued)

TABLE 13.6 *FIFA 2008* Visual Settings combinatorial table (continued)

	Camera	Radar	TS Display	HUD	Indicator
12	DynamicV2	OFF	ON	NameBar	Number
13	Tele	2D	OFF	NameBar	Name
14	Tele	3D	ON	Indicator	Number
15	EndToEnd	OFF	OFF	Indicator	Number
16	Pro	OFF	ON	NameBar	Name

NOTE *The files for all of the FIFA combinatorial update examples are included with the book's companion files, which can be obtained by writing to info@merclearning.com.*

FIFA 2009 made only one small addition to the visual settings: a "Broadcast" choice for the Camera option. This is easily incorporated into an additional row in the Allpairs input file shown in Listing 13.2, which produces the tests shown in Table 13.7. Adding the "Broadcast" value was inexpensive, costing only three more tests.

LISTING 13.2 Allpairs File for FIFA 2009 Visual Settings Changes

Camera	Radar	TS-Display	HUD	Indicator
Dynamic	2D	OFF	NameBar	Name
DynamicV2	3D	ON	Indicator	Number
Tele	OFF			
EndToEnd				
Pro				
Broadcast				

TABLE 13.7 *FIFA 2009* Visual Settings combinatorial table

	Camera	Radar	TS Display	HUD	Indicator
1	Dynamic	2D	OFF	NameBar	Name
2	Dynamic	3D	ON	Indicator	Number
3	DynamicV2	2D	ON	NameBar	Number
4	DynamicV2	3D	OFF	Indicator	Name
5	Tele	OFF	OFF	NameBar	Number
6	Tele	OFF	ON	Indicator	Name
7	EndToEnd	2D	OFF	Indicator	Number
8	EndToEnd	3D	ON	NameBar	Name
9	Pro	2D	ON	Indicator	Name
10	Pro	3D	OFF	NameBar	Number
11	Broadcast	OFF	OFF	Indicator	Name
12	Broadcast	2D	ON	NameBar	Number
13	Dynamic	OFF	ON	NameBar	Number
14	DynamicV2	OFF	OFF	NameBar	Name
15	Tele	2D	OFF	Indicator	Number
16	Tele	3D	ON	NameBar	Name
17	EndToEnd	OFF	ON	Indicator	Number
18	Pro	OFF	OFF	NameBar	Name
19	Broadcast	3D	OFF	Indicator	Number

Because only the Camera parameter list was expanded for *FIFA 09*, an alternative to producing an entirely new table would be to add additional rows to the table, which will cover the combinations necessary for the added "Broadcast" choice. The advantage of doing this is that you can continue to use the tests you have already produced so you do not have to update your test management system or re-do the automation for those tests. The test suite will grow based on which parameter acquired the additional value. In this case, the new Camera parameter added only three more tests because that was the highest number of choices for any of the other parameters. To form the required pairs, the new choice has to be combined with each of the choices for the other parameters in the test. If a new value was added to the HUD instead of the Camera, then five new tests would be required to pair the new HUD value with each of the five Camera choices: "Dynamic," "DynamicV2," "Tele," "EndToEnd," and "Pro." Table 13.8 shows test cases 17-19 added on to the end of the *FIFA 2008* table.

TABLE 13.8 *FIFA 2009* Visual Settings appended to the *FIFA 2008* table

	Camera	Radar	TS Display	HUD	Indicator
1	Dynamic	2D	OFF	NameBar	Name
2	Dynamic	3D	ON	Indicator	Number
3	DynamicV2	2D	ON	NameBar	Number
4	DynamicV2	3D	OFF	Indicator	Name
5	Tele	OFF	OFF	NameBar	Number
6	Tele	OFF	ON	Indicator	Name
7	EndToEnd	2D	OFF	Indicator	Number
8	EndToEnd	3D	ON	NameBar	Name
9	Pro	2D	ON	Indicator	Name
10	Pro	3D	OFF	NameBar	Number
11	Dynamic	OFF	OFF	Indicator	Name
12	DynamicV2	OFF	ON	NameBar	Number
13	Tele	2D	OFF	NameBar	Name
14	Tele	3D	ON	Indicator	Number
15	EndToEnd	OFF	OFF	Indicator	Number
16	Pro	OFF	ON	NameBar	Name
17	**Broadcast**	**2D**	**OFF**	**NameBar**	**Name**
18	**Broadcast**	**3D**	**ON**	**Indicator**	**Number**
19	**Broadcast**	**OFF**	**OFF**	**NameBar**	**Name**

FIFA 10 retained the same choices as *FIFA 09*, so you get to keep those tests without any changes. Now let's take a look at what changed in *FIFA 11*.

First, the Indicator visual setting was renamed "Player Indicator," and a Gamertag Indicator On/Off setting was added. Second, a Net Tension setting was added, allowing you to choose between "Default," "Regular," "Loose," or "Tight" tension. Finally, the "DynamicV2" Camera setting was renamed "Co-Op." When you update the Allpairs input file, make sure each row has a total of seven columns (six tabs) to account for the two added parameters. Listing 13.3 shows the Allpairs input data for the *FIFA 11* visual settings.

LISTING 13.3 Allpairs File for FIFA 2011 Visual Settings Changes

Camera	Radar	TS-Display	HUD	Player Indicator	Gamertag Indicator	Net Tension
Dynamic	2D	OFF	NameBar	Name	OFF	Default
Co-Op	3D	ON	Indicator	Number	ON	Regular
Tele	OFF					Loose
EndToEnd						Tight
Pro						
Broadcast						

This time the change has a more significant impact on the test suite. Table 13.9 shows that six more tests are required, so now the table has grown to more than double the size of that which was first produced for *FIFA 2007*. You also need to consider that the time to setup, run, automate, and check the results of these tests has become more complicated because of the additional parameters that have to be accounted for.

TABLE 13.9 *FIFA 2011* Visual Settings combinatorial table

	Camera	Radar	TS-Display	HUD	Player Indicator	Gamertag Indicator	Net Tension
1	Dynamic	2D	OFF	NameBar	Name	OFF	Default
2	Dynamic	3D	ON	Indicator	Number	ON	Regular
3	Co-Op	3D	ON	NameBar	Number	OFF	Default

(Continued)

TABLE 13.9 *FIFA 2011* Visual Settings combinatorial table (continued)

	Camera	Radar	TS-Display	HUD	Player Indicator	Gamertag Indicator	Net Tension
4	Co-Op	2D	OFF	Indicator	Name	ON	Regular
5	Tele	OFF	ON	NameBar	Name	ON	Loose
6	Tele	OFF	OFF	Indicator	Number	OFF	Tight
7	EndToEnd	2D	ON	Indicator	Number	OFF	Loose
8	EndToEnd	3D	OFF	NameBar	Name	ON	Tight
9	Pro	OFF	ON	Indicator	Name	ON	Default
10	Pro	2D	OFF	NameBar	Number	OFF	Regular
11	Broadcast	3D	OFF	Indicator	Name	OFF	Loose
12	Broadcast	2D	ON	NameBar	Number	ON	Tight
13	Dynamic	OFF	ON	NameBar	Name	OFF	Regular
14	Co-Op	OFF	OFF	NameBar	Number	ON	Loose
15	Tele	2D	OFF	Indicator	Number	ON	Default
16	Tele	3D	ON	NameBar	Name	OFF	Regular
17	EndToEnd	OFF	OFF	Indicator	Number	OFF	Default
18	Pro	3D	ON	Indicator	Name	OFF	Tight
19	Broadcast	OFF	ON	NameBar	Name	ON	Default
20	Dynamic	~3D	OFF	Indicator	Number	ON	Loose
21	Co-Op	~2D	ON	Indicator	Name	OFF	Tight
22	EndToEnd	~OFF	ON	NameBar	Name	ON	Regular
23	Pro	~2D	OFF	NameBar	Number	ON	Loose
24	Broadcast	~3D	OFF	Indicator	Number	OFF	Regular
25	Dynamic	~OFF	OFF	NameBar	Number	ON	Tight

It is very expensive to run all of your tests for every incremental change to a game, so choosing the right set of tests to run can make a big difference in how fast your team can retest and re-certify the new code. Good regression testing is a combination of safety, history, and intuition. The right formula will balance higher quality with a shorter time frame for getting your product out to your avid players.

Tests which are constructed in a consistent and rational manner make it easier for testers to maintain, update, and execute their tests efficiently. Like well-built durable objects in the real world, reusable tests should last a long time, be useful in many situations, and require little or no maintenance to continue functioning.

EXERCISES

6. Create a bug-fix TFD for the following *Gears of War 2* issue: *"An issue where players couldn't chainsaw enemy meatshields if the meatshields were already damaged."*
7. Use a 2-bubble TFD pattern to provide tests for at least three scenarios you might find in a vampire role-playing game. Consider situations from both the vampire's perspective and the perspective of other characters who might appear in the game.
8. The gameplay video at *https://youtu.be/ia-A6qBA-qk* contains video of a match played by the two teams shown in Figure 13.3. Write down all of the situations (including timecodes) you can find on the video where the left center midfielder's (LCM) name should appear but is not shown.
9. Update Table 13.9 to account for a "3DTV" Gamertag Indicator value, adding the minimum number of new test cases needed to maintain full pairwise coverage.

REFERENCE

Fandom.com. n.d. "Gears of War 2 Title Updates." Accessed January 14, 2024. *https://gearsofwar.fandom.com/wiki/Gears_of_War_2_Title_Updates*.

CHAPTER 14

TESTING BY THE NUMBERS

Product metrics, such as the number of defects found per line of code, tell you how fit the game code is for release. Test metrics can tell you about the effectiveness and efficiency of your testing activities and results. A few pieces of basic test data can be combined in ways that reveal important information that you can use to keep testing on track, while getting the most out of your tests and testers.

TESTING PROGRESS

Collecting data is important to understanding where the test team is and where they are headed in terms of meeting the needs and expectations of the overall game project. Data and charts can be collected by the test lead or the individual testers. Take responsibility for knowing how well you and your team are doing. For example, in order to estimate the duration of the test execution for any portion of the game project, estimate the total number of tests to be performed. This number is combined with data on how many tests can be completed per staff-day of effort, how much of a tester's calendar time is actually spent on testing activities, and how many tests you can expect to be redone.

Figure 14.1 provides a set of data for a test team starting to run tests against a new code release. The project manager worked with the test lead to use an estimate of 12 tests per day as the basis for projecting how long it would take to complete the testing for this release.

Date	Daily Execution		Total Execution	
	Planned	Actual	Planned	Actual
22-Dec	12	13	12	13
23-Dec	12	11	24	24
28-Dec	12	11	36	35
29-Dec	12	12	48	47
30-Dec	12	8	60	55
4-Jan	12	11	72	66
5-Jan	12	10	84	76
6-Jan	12	11	96	87
7-Jan	12	11	108	98
8-Jan	12	16	120	114
10-Jan	12	10	132	124
11-Jan	12	3	144	127
12-Jan	12	7	156	134

FIGURE 14.1 Planned and actual test execution progress data

Thirteen days into the testing, the progress lagged what had been projected, as shown in Figure 14.2. It looks like progress started to slow on the fifth day, but the team was optimistic that they could catch up. By the tenth day, they seemed to have managed to steer back toward the goal, but during the last three days the team lost ground again, despite the reassignment of some people onto and off of the team.

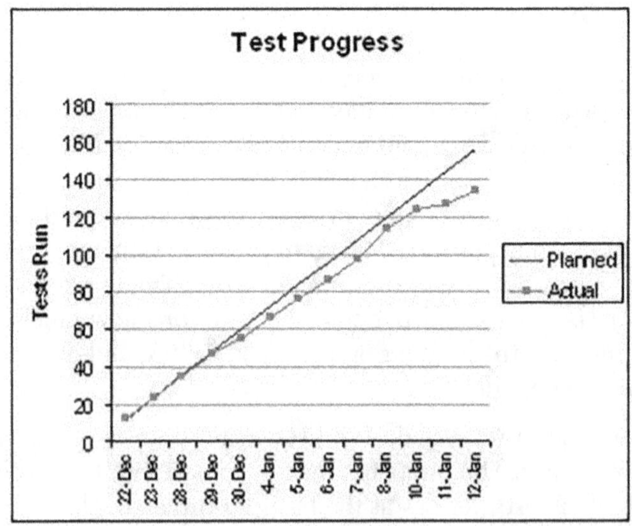

FIGURE 14.2 Planned and actual test execution progress graph

To understand what is happening here, data was collected for each day that a tester was available to do testing, and the number of tests he or she completed each day. This information can be put into a chart, as shown in Figure 14.3. The totals show that an individual tester completes an average of about four tests a day.

DATE	TESTER					TESTER DAYS	COMPLETED TESTS
	B	C	D	K	Z		
22-Dec	*				*	2	13
23-Dec	*				*	2	11
28-Dec	*				*	2	11
29-Dec	*				*	2	12
30-Dec	*				*	2	8
4-Jan	*		*		*	3	11
5-Jan	*		*		*	3	10
6-Jan	*		*		*	3	11
7-Jan	*		*		*	3	11
8-Jan		*	*	*	*	4	16
10-Jan	*	*	*			3	10
11-Jan	*	*	*			3	3
12-Jan			*			1	7
					TOTALS	33	134
					TESTS/TESTER DAY		4.06

FIGURE 14.3 Test completion rate per tester per day

Once you have the test effort data for each person and each day, you must compare the test effort people have contributed to the number of work days they were assigned to participate in system testing. Ideally, this ratio would come out to 1.00. The numbers you actually collect will give you a measurement of something you felt was true but could not prove before: most testers are unable to spend 100% of their time on testing. This being the case, do not plan on testers spending 100% of their time on a single task! Measurements will show you how much to expect from system testers, based on various levels of participation. Some testers will be dedicated to testing as their only assignment. Others perhaps perform dual role, such as developer/tester or QA engineer/tester. Collect effort data for your team members that fall into each category, as shown in Figure 14.4.

FULL-TIME TESTERS

WEEK	1	2	3	4	5	6	7	TOTAL
TESTER DAYS	15.5	21.5	35.5	31.5	36.5	22	23.5	186
ASSIGNED DAYS	44	50	51	53	50	41	41	330
							FULL-TIME TESTER AVAILABILITY	56%

PART-TIME TESTERS

WEEK	1	2	3	4	5	6	7	TOTAL
TESTER DAYS	0	0	0	18.5	18.5	6	15	58
ASSIGNED DAYS	0	0	0	49	54	53	46	202
							PART-TIME TESTER AVAILABILITY	29%

CUMULATIVE

TESTER DAYS	244
ASSIGNED DAYS	532
AVAILABILITY	46%

FIGURE 14.4 Tester participation rate calculations

These data lead to a number of important points. One is that, given tester "overhead" tasks such as training, meetings, preparing for demos, and so on, a full-time tester might be able to contribute only about 75% of his or her time at best, and 50% to 60% on average, over the course of a long project. If you are counting on people with other responsibilities—for example, artists, developers, or designers—to help with testing, then expect only half a much participation from them as the full-time testers. Using the numbers in Figure 14.4, that would be about 30% of their total available time. You will need to make these measures for your own particular project.

By combining the individual productivity numbers to find a team productivity number, you can see that this team performs only half as many tests as they could if they had 100% of their time to perform testing. This number can be combined with your effort estimate to give an accurate count of calendar work days remaining before testing will be finished. Using the number of 125 tests remaining and a staff size of 11 testers, you would approximate that 11 staff-days of testing remain. Now that you know what the team's actual productivity is, however, you divide 11 by 46%, resulting in 24 calendar work days remaining, or nearly five "normal" work weeks. If you had committed to the original, optimistic number of 11 days, there would have been much trouble when the tests were not actually completed until three weeks after they had been promised!

You need this kind of information to answer questions such as "How many people do you need to get testing done by Friday?" or "If I can get you two more testers, when can you be done?"

TIP

Steady Progress Wins the Race
Remember that it is easier to stay on track by getting a little extra done day-by-day than by trying to make up a large amount of time later in a panic situation.

Going back to Figure 14.1, you can see that on 8-Jan the team was only six tests behind the goal. Completing one extra test on each of the previous six work days would have kept the team on goal. If you can keep short-term commitments to stay on track, you will be able to keep the long-term commitment to deliver completed testing on schedule.

TESTING EFFECTIVENESS

Measure your Testing Effectiveness (TE) by adding up defects and dividing by the number of tests completed. This measurement tells you not only how "good" the current release is compared to previous ones, it can also be used to predict how many defects will be found by the remaining tests for that release. For example, with 30 tests remaining and a TE of 0.06, testers should find approximately two more defects. This could be a sign to developers to delay a new code release until the two expected defects are identified, classified, and removed. An example table of TE measurements is shown in Figure 14.5.

Code	Defects		Tests Run		Defects/Test	
Release	New	Total	Release	Total	Release	Total
DEV1	34	34	570	570	0.060	0.060
DEV2	47	81	1230	1800	0.038	0.045
DEV3	39	120	890	2690	0.044	0.045
DEMO1	18	138	490	3180	0.037	0.043
ALPHA1	6	144	220	3400	0.027	0.042

FIGURE 14.5 Test Effectiveness measurements

You should measure TE for each release as well as for the overall project. Figure 14.6 shows a graphical view of this TE data.

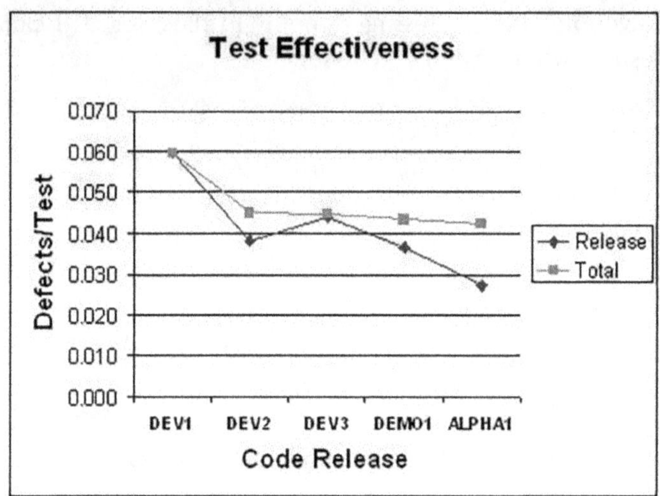

FIGURE 14.6 Test Effectiveness graph

Notice how the cumulative TE reduced with each release and settled at .042. You can take this measurement one step further by using test completion and defect detection data for each tester in order to calculate individual TEs. Figure 14.7 shows a snapshot of tester TEs for the overall project. You can also calculate each tester's TE per release.

TESTER	B	C	D	K	Z	TOTAL
TESTS RUN	151	71	79	100	169	570
DEFECTS FOUND	9	7	6	3	9	34
DEFECTS/TEST	0.060	0.099	0.076	0.030	0.053	0.060

FIGURE 14.7 Test Effectiveness measured for individual testers

Note that for this project, the effectiveness of each tester ranges from 0.030 to 0.099, with an average of 0.060. The effectiveness is perhaps as much a function of the particular tests each tester was asked to perform as it is a measure of the skill of each tester. Like the overall TE measurement, however, this number can be used to predict how many additional defects a particular tester could find when performing a known number of tests. For example, if Tester C has 40 more tests to perform, expect her to find about four more defects.

In addition to measuring how many defects you detect (quantitative analysis), it is important to understand the severity of defects introduce with each release (qualitative analysis). Using a defect severity scale of 1 to 4, where 1 is the highest severity, detection of new severity 1 and 2 defects should be reduced to 0 prior to shipping the game. Severity 3 and 4 defect detection should be on a downward trend approaching 0. Figure 14.8 provides examples of severity data.

Release	Defects by Severity				
	1	2	3	4	All
Dev1	7	13	13	1	34
Dev2	4	11	30	2	47
Dev3	2	3	34	0	39
Demo1	1	2	12	3	18
Alpha1	0	0	6	0	6

FIGURE 14.8 Defect severity trend data

Figure 14.9 graphs the trend of the severity data listed in Figure 14.8. Take a moment to examine the graph. What do you see?

Notice that the severity 3 defects dominate. They care also the only category to significantly increase after "Dev1" testing, except for some extra 4s appearing in the "Demo1" release. When you set a goal that does not allow any severity 2 defects to be in the shipping game, there will be a tendency to push any borderline severity 2 issues into the severity 3 category. Another explanation could be that the developers focus their efforts on the 1s and 2s so they leave the 3s alone early in the project, with the intention of dealing with them later. This approach bears itself out in Figures 14.8 and 14.9, where the severity 3 defects are brought way down for the "Demo1" release and continue to drop in the "Alpha1" release. Once you see "what" is happening, try to understand "why" it is happening that way.

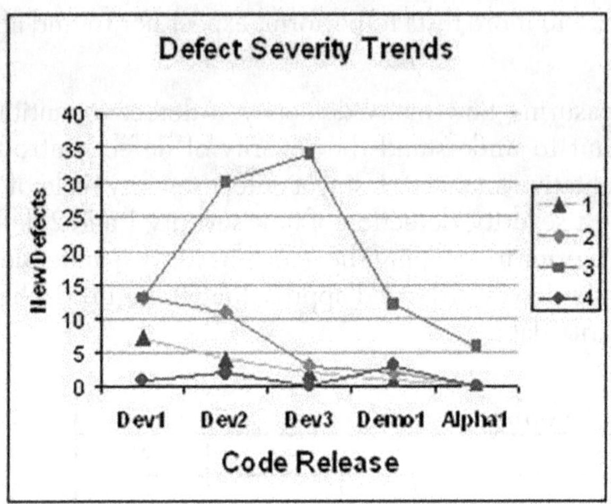

FIGURE 14.9 Defect severity trend graph

TESTER PERFORMANCE

You can implement some other measurements to encourage testers to find defects and to give them a sense of pride in their skills. One of them is a *Star Chart*. This chart is posted in the testing area and shows the accomplishments of each tester according to how many defects they find of each severity. Tester names are listed down one side of the chart and each defect is indicated by a stick-on star. The star's color indicates the defect's severity. For example, you can use blue for 1, red for 2, yellow for 3, and silver for 4. Points can also be assigned to each severity (for example, A = 10, B = 5, C = 3, D = 1), and a "Testing Star" can be declared at the end of the project based on who has the point points.

NOTE

In our experience, this chart has led to a sense of friendly competition among testers, increased their determination to find defects, promoted tester ownership of defects, and has caused testers to pay more attention to their own productivity. This approach turns testing into a game for the testers to play while they are testing the games.

Figure 14.10 shows what a Star Chart can look like prior to applying the tester's stars.

STAR CHART FOR XYZZY	
TESTERS	STARS (Sev. 1 = BLUE, 2 = RED, 3 = YELLOW, 4 = SILVER)
B	
C	
D	
K	
Z	

FIGURE 14.10 Empty Star Chart

If you are concerned about testers getting into battles over defects and not finishing their assigned tests quickly enough, you can create a composite measure of each tester's contribution to text execution and defects found. Add the total number of test defects found and calculate a percentage for each tester, based on how many they found divided by the project total. Then do the same for tests run. You can add these two numbers for each tester. Whoever has the highest total is the "Best Tester" for the project. This might or might not turn out to be the same person who becomes the Testing Star. Here is how this works for Testers B, C, D, K, and Z for the "Dev1" release:

- Tester B executed 151 of the team's 570 "Dev1" tests. This comes out to 26.5%. B has also found 9 of the 34 "Dev1" defects, which is also 26.5%. B's composite rating is 53.
- Tester C ran 71 of the 570 tests, which is 12.5%. C found 7 of the 34 total defects in "Dev1," which is 20.5%. C's rating is 33.
- Tester D ran 79 tests, which is approximately 14% of the total. D also found 6 defects, which is about 17.5% of the total. D earns a rating of 31.5.
- Tester K ran 100 tests and found 3 defects. These represent 17.5% of the total and about 9% of the defect total. K has a 26.5 rating.
- Tester Z ran 169 tests, which is about 29.5% of the 570 total. Z found 9 defects, which is 26.5% of that total. Z's total rating is 56.
- Tester Z has earned the title of "Best Tester."

Be careful to use this system for good and not for evil. Running more tests or claiming credit for new defects should not come at the expense of other people or the good of the overall project. You could add factors to give more weight to higher-severity defects to discourage testers from spending all their time chasing and reporting low-severity defects that will not contribute as much to the game as a few very important high-severity defects.

Use this system to encourage and exhibit positive test behaviors. Remind your team (and yourself!) that some time spent automating tests could have generous payback in terms of test execution. Likewise, spending a little time up front to effectively design your tests, before you run off to start banging on the game controller, will probably lead you to more defects.

This chapter introduced you to a number of metrics you can collect to track and improve testing results. Each metric from this chapter is listed below, together with the raw data you need to collect for each, mentioned in parentheses:

- Test Progress Chart (# of tests completed by team each day, # of tests required each day)
- Test Completed/Days of Effort (# of tests completed, # days of test effort for each tester)
- Test Participation (# of days of effort for each tester, # of days each tester assigned to test)
- Test Effectiveness (# of defects, # of tests for each release and/or tester)
- Defect Severity Profile (# of defects of each severity for each release)
- Star Chart (# of defects of each severity for each tester)
- Testing Star (# of defects of each severity for each tester, point value of each severity)
- Best Tester (# of tests per tester, # of total tests, # of defects per tester, # of total defects)

Testers or test leads can use these metrics to aid in planning, predicting, and performing game testing activities.

EXERCISES

1. How does the data in Figure 14.3 explain what is happening on the graph in Figure 14.2?
2. How many testers do you need to add to the project represented in Figures 14.1 and 14.2 to bring the test execution back on plan in the next 10 working days? The testers will begin work on the very next day that is plotted on the graph.
3. Tester C has the best TE, as shown in Figure 14.7, but did not turn out to be the "Best Tester." Explain how this happened.
4. You are Tester X working on the project represented in Figure 14.7. If you have run 130 tests, how many defects did you need to find in order to become the "Best Tester?"
5. Describe three positive and three negative aspects of measuring the participation and effectiveness of individual testers. Do not include any aspects already discussed in this chapter.

CHAPTER 15

Software Quality

Software quality can be determined by how well the product performs the functions of which it was intended. For game software, this includes the quality of the player's experience plus how well the game features are implemented. Various activities can be performed to evaluate, measure, and improve game quality.

In the book *Quality is Free: The Art of Making Quality Certain* (Crosby 1980), Philip Crosby states that "Quality is free." This should be the high concept of your quality program. If the cost of performing some quality function is not expected to produce an eventual savings, find a way to do it cheaper or better. If you cannot, then stop doing it.

Another way of considering the *process* (separate from the *product*) of both game development and game testing was best expressed by longtime UCLA men's basketball coach John Wooden: "If you don't have the time to do it right, when will you have the time to do it over?" Too often, under the intense deadline pressure of game development, team members may sacrifice adhering to standards because they think they do not have time to "do it right." The truth is that "doing it right" will ultimately save them and their team valuable time on this project, or the next, or in the months and years to come in the case of a "live service" game.

GAME QUALITY FACTORS

Different players may have different criteria for what makes a game "good" for them. Some qualities are likely to be important to many game customers:

- quality of the story
- quality of the game mechanics
- quality (for example, style and realism) of in-game audio and visual effects
- quality of the download and update experience
- beauty of the visual style
- use of humor and exaggeration
- "human-like" non-player character Artificial Intelligence (AI)

Games should also have an interface that is easy to use and clear to understand. This includes both the graphical user interface elements presented on the screen during gameplay and the game control(s) provided for the player to operate and to affect the game. The user interface can consist of multiple elements such as on-screen displays and menus. The game control includes the way players control and operate their characters (or teams, cars, and armies) during the game, as well as the way they can control their experience through point-of-view and lighting settings. The game should also support a variety of controllers that are especially suited to the game's genre, such as joysticks for air combat, guitars for making music, and steering wheels for driving.

Another factor in providing a quality experience for the user is to ensure game code and assets are compatible with the memory constraints of the target platform. This includes the available working memory required for the game to run properly as well as the size, quantity and types of target media supported such as CD-ROMs, DVDs, digital downloads, or virtual reality content.

Higher memory requirements may affect game performance while time is spent switching game assets in and out of memory during play. The impact is magnified when assets are sent from a remote server to the console, PC, or mobile device. If the game code and assets do not fit within the memory footprint of the least expensive device, the market for the game and its profit potential are reduced.

Handheld device and console memory is not as upgradable as PC memory is. Games have to fit within the memory constraints of the onboard chips, removable memory, and hard drive devices supported by the target platform. Mobile games are the most constrained in terms of available fixed and removable memory and tend to use up more memory as fixes and updates are applied over the life of the game. Both mobile and console players are likely to reach a point where their memory consumption reaches the limitations of their device and have to make a decision about deleting a less-frequently played game for the exciting new download that caught their eye.

Any efforts at code *crunching*—optimizing code or data so that the game runs more efficiently on the target hardware platform(s)—get more expensive the later they happen in the development cycle. The cost is not just in the labor to do the reduction work. Shrinking game code or reformatting assets to fit on the target media or memory footprint can introduce new, hard-to-find bugs late in the project. This creates an extra burden on development, project management, defect tracking, version control, and testing.

GAME QUALITY APPRAISAL

The actual quality of a game is established by its design and subsequent implementation in code. However, the appraisal activities are necessary to identify the difference between what was produced and what should have been produced. Once identified, these differences can be repaired before (and sometimes after) releasing the game.

Testing is considered an appraisal activity. It establishes whether the game code performs the functions for which it was intended. However, testing is not the most economical way to find game defects. It is better to catch problems at the point they are introduced.

Having peers review game deliverables as they are being produced provides immediate feedback and the opportunity to repair problems before they are integrated into the rest of the game. It will be much harder and more expensive to find and repair these problems at later phases of the project.

Peer reviews come in different "flavors." In each case, there are times when you, the tester, may be required to participate. If you do not put in the necessary time and effort to contribute to the review, you and your team will be less likely to be asked to participate in the future. Make sure you take this responsibility seriously when it is your turn to participate in a review.

Walkthroughs

Walkthroughs are one form of peer review. A general outline of a walkthrough might be:

1. The leader (for example, the lead designer) secures a room and schedules the walkthrough.
2. The leader begins the meeting with an overview of work including scope, purpose, and special considerations.
3. The leader displays and presents document text and diagrams.
4. Participants ask questions and raise issues.
5. New issues raised during the walkthrough are recorded during the meeting.

The room should comfortably fit the number of people attending and have a projector for presentations. A whiteboard or paper easel pad can be used by the leader or participants to elaborate on questions or answers. Limit attendance to around six to eight people. This should not turn into a team meeting. Only include a representative from each project role that is potentially affected by the work you are walking through. For example, someone from the art team does not have to be in most code design walkthroughs, but there should be an experienced game artist there when graphics systems designs are being discussed. Do not invite the test lead to every single walkthrough that affects the test team. If you do, then game knowledge and walkthrough experience may not get passed on to other testers. This also keeps the test lead from spending too much time on walkthroughs and not enough time on test leading. Work with the test lead to find other capable representatives on her team. If you are the test lead, send someone capable from your team in your place when you can.

Be sure to invite one or more developers to your test walkthroughs. It is a great way to find out if what you intend to test is really what the game is going to do once it is developed. Conversely, get yourself invited to design and code walkthroughs. Learn about the design techniques and programming language your team is using. Even if you do not have any comments to improve the author's work, you can use what you learn there to make your tests better.

It is also not a bad idea to use some walkthroughs as mentoring or growth opportunities for people on your team. The "guests" should limit their own questions and comments during the meeting to the material being presented and have a follow-up time with their "host" to go over any other questions about procedures, the design methodology being used, and so on. This probably should not be done for every walkthrough, but only in situations where someone already has a background in the topic or is expected to grow into a lead role for some portion of the project.

Here is a list of representatives to consider inviting to walkthroughs of various project artifacts:

- **Technical Design Document (TDD):** tech lead, art director, producer, project manager
- **Storyboard:** producer, dev lead, artists
- **Software Quality Assurance Plan (SQAP):** Project manager, producer, development lead, test lead, QA lead, and engineer(s)
- **Code designs, graphics:** key developers, art representative, test representative
- **Code designs, other:** key developers, test representative
- **Code:** key developers, key testers
- **Test plan:** project manager, producer, development lead, key testers
- **Tests:** - feature developer, key testers

Relevant topics to cover in walkthroughs include

- possible implementations
- interactions
- appropriate scope
- traceability to earlier work products
- completeness

Issues raised during the walkthrough are also recorded during the meeting. Sometimes the presenter will realize a mistake simply by talking about his work. The walkthrough provides an outlet for that. One participant acts as a recorder, recording issues and presentation points that are essential to understand the material. Other participants may end up using the information for downstream activities, such as coding or testing. The leader is responsible for promptly closing each issue and distributing the meeting notes to the team within one week of the walkthrough. QA is expected to follow up by checking that the issues were indeed closed before any work was done based on the material that was walked through and that the notes were distributed to the participants.

Reviews

Reviews are a little more intimate than walkthroughs. Reviewers are expected to prepare their comments prior to the review meeting and submit them to the review leader so that they can be consolidated prior to the actual meeting. Comments sent electronically are easier to compile and understand than handwritten comments. Be sure to let the review leader know when you are going to submit a pen-and-paper mark-up instead of an electronic file. The review leader may or may not be the author of the material being reviewed.

The review itself can be an in-person meeting between the author and reviewers or simply a review of the comments by the author alone, who then contacts individual reviewers if she has any questions about their issues. An in-between approach is for the author to look over the reviewer comments prior to the review meeting and to limit the meeting to discussions over the few issues that the author disagrees with or has questions about. This meeting can also take place virtually, which is especially useful for projects distributed across studios that are separated by space and time.

During the meeting, someone (usually the review leader) must take notes and publish the resolution of each item to the team. If the opinions of a reviewer differ from what the author believes should be done, decisions on technical matters are left to the author, whereas procedural matters can be resolved by QA.

Checklist-based Reviews

Another form of review takes place between only two people: the author and a reviewer. In this case, the reviewer follows a checklist to look for mistakes or omissions in the author's work. The checklist should be thorough and based on specific mistakes that are common for the type of work being reviewed. Requirements, code, and test reviews of this type would each use different checklists. At times, it would even be appropriate to have checklists specific to a game project. These checklists should constantly evolve to include new types of mistakes that start to show up. Mistakes found during the checklist review that were not on the checklist should be recorded and considered for use in the next version. Technology, personnel, and methodology changes could all lead to new items being added to the checklist.

Inspections

Inspections are more structured than reviews. Fagan Inspections are one particular inspection methodology from which many others have been derived. They were defined by Michael Fagan in the 1970s based on his work at IBM, and are now part of the Fagan Defect-Free Process (Fagan 86, 744).

A Fagan Inspection follows these steps:

1. Planning
2. Overview
3. Preparation
4. Meeting
5. Rework
6. Follow-up
7. Casual Analysis

The inspection meeting is limited to four people, which each session taking no more than two hours. Larger work should be broken up into multiple sessions. These guidelines are based on data that shows a decline in the effectiveness of the inspection if these limits are exceeded. If you do not know your inspection rates, such as pages per hour or lines of code per hour, measure them for the first 10 or so inspections you do. Then use those results to calculate how many sessions are needed for any future inspections.

In the Fagan Inspection method, each participant plays a specific role in the inspection of the material. The *Moderator*, who is not the author, organizes the inspection and checks that the materials to be inspected satisfy predefined criteria. As with the checklist reviews, you will need to establish these criteria for different items that you will be inspecting. Once the criteria are met, the Moderator schedules the review meeting, plus an "overview" session that takes place prior to the review. This is to discuss the scope and intent of the inspection with the participants. Participants may also have questions that can be answered here or soon after the meeting. Typically, there should be two working days between the overview and the inspection meeting. This is to give reviewers adequate preparation time. Each of the inspectors is assigned a role to play in the inspection meeting. The *Reader* is expected to paraphrase the material being inspected. The idea is to communicate any implied information or behavior that the Reader interprets to see if it matches the Author's intended function. For example, here is a line of code to read:

```
LoadLevel(level[17]], highRes, 0);
```

You could just say "Call `LoadLevel` with level seventeen, high res and zero." A better reading for inspection purposes would be say "Call `LoadLevel` without checking the return value. Pass the level information using a constant index of seventeen, the stored value of `highRes` and a hard-coded zero." This second reading raises the following potential issues:

1. The return value of `LoadLevel` is not checked. Should it return a value to indicate success, or a level number to verify that the level you intended to load did get loaded?
2. Using a constant index for the level number may not be a good practice. Should the level number come from a value passed to the routine that this code belongs to, or should the number 17 be referenced by a more descriptive defined constant such as `HAIKUDUNGEON` in case something in the future causes the level numbering to be reordered?
3. The value of 0 provides no explanation about its function or the parameter it is being assigned to.

You can get similar results from reading test cases. Having another person try to understand the literal meaning of each of your test steps word for word may not turn out as you intended.

The *Tester* does not have to be the person from the test team. This is a role where the person questions things like whether the material being inspected is internally consistent or consistent with any project documents it is based on. It is also good if the Tester can foresee how this material will fit in with the rest of the project and how it would potentially be tested.

A *Recorder* takes detailed notes about the issues raised in the inspection. The Recorder is a second role that can be taken on by any of the four people involved. The Reader is probably not the best choice for Recorder and you may find that it works best if the Moderator accepts the Recorder role. The Moderator also keeps the meeting on track by limiting discussions to the material at hand.

Throughout the meeting, the participants should not feel confined by their roles. They need to become engaged in discussions of potential issues or how they interpret material. A successful inspection is one that invites the *Phantom Inspector*. This is neither an actual person nor a supernatural manifestation. Rather, it is a term to explain the source of extra issues that are raised by the inspection team coming together and feeding off each other's roles.

Once the meeting has concluded, the Moderator determines whether any rework is required before the material can be accepted. He continues to work with the Author to follow up on issues until they are closed. An additional inspection may be necessary, based on the volume or complexity of the changes.

The final step of this process involves casual analysis of the product (inspected item) faults and any inspection process (overview, preparation, meeting, and so on) problems. Issues can be discussed, such as how the overview could have been more helpful, or requiring stricter compiler flags to be set that could flag certain code defects prior to submitting the code for inspection.

GAME STANDARDS

Among its many responsibilities, the QA team should establish that the project work products follow the right formats. This includes assuring that the game complies with any standards that apply. User interface standards and coding standards are two kinds of standards applicable to game software.

User Interface Standards

User Interface (UI) *standards* help players identify with your game. Below are some examples of UI standards, which are derived from a 2004 Game Developers Conference presentation, "Cross-Platform User Interface Development," by Rob Caminos and Tim Stellmach (Caminos 2004). As part of your QA function, you would examine relevant screens to confirm they had the properties and characteristics called for in these standards:

1. Text should be large and thick, even at the expense of creating an extra page of text.
2. Make all letter characters the same size.
3. Avoid using lowercase letters. Instead, use smaller versions of uppercase letters.
4. Use an outline for the font where possible.
5. On-screen keyboards should resemble the look of an actual keyboard.
6. On-screen keyboards should have the letters arranged alphabetically. Do not use the QWERTY arrangement.
7. Split alphabet, symbol, and accent characters into three separate on-screen keyboards.
8. Common functions such as Done, Space, Backspace, Caps Lock, and switching between character sets should be mapped to available buttons on the game controller.
9. Assign Space and Backspace keyboard functions to the left and right shoulder buttons.
10. Each menu should fit on one screen.
11. The cursor should blatantly draw attention to the currently selected menu item.
12. Avoid horizontal menus.
13. Vertical menus should consist of no more than 6-8 items, each with its own button.
14. Menus should be cyclic, allowing the player to loop through the menu choices.

15. Leave extra room for text localization. (Some languages, such as German, may require more letters per word than your game's native language.)

16. Place button icons next to their functions instead of using lines to connect the functions to the buttons.

17. Point button icons to their location on the controller.

18. Separate thumb-stick movement functions from button functions.

Additional standards could apply to consistent keyboard assignments ("F1 should always be the Help button") or the flexibility of game controller options ("There should always be an option to enable or disable vibration").

Your list of standards can be used as a checklist that gets filled out for each screen. The checklist should include other information such as the QA person's name, the date of the appraisal, the name of the software build or identifier being checked, and the name of the menu screen. Do not wait until the UI is coded and put into a release before you check it. Work with developers to verify that the standard is being followed in their UI design. Some checking should also take place after code is released to verify that the implementation matches the intent. This may include a suite of tests that specifically check that each UI standard is met.

You may find that some of these items above make perfect sense for your game, while some do not. Use what is right for you and your players. The important thing is to have some standards, have a reason for including each item in the standard, and have a way to check periodically that the team uses the standard.

Coding Standards

Coding standards can prevent the introduction of defects when the game code is written. Some of the topics typically addressed by coding standards include:

- file naming conventions
- header files
- comment and indentation styles
- use of macros and constants
- use of global variables

To some critics, coding standards pay too much attention to the format of the code rather than its substance. However, there must be some reason why development tool companies continue to provide more coding assistance using visual means such as colors and graphs. Both have the same goal in mind: to help the developer get the code right the first time.

Even so, coding standards are not just about formatting. Many of the rules are designed to address important issues such as portability, clarity, modularity, and reusability. The importance of these standards is magnified in a project that is distributed across different teams, sites, and countries. There are few things less fun than tracking down a defect caused by one team defining `SUCCESS` as 0 and another team defining `SUCCESS` as 1.

Here are some examples of the C++ coding standards used at developer Ronimo Games (van Dongen 2017):

- Every comma is followed by a space, for example `doStuff(5, 7, 8)`.
- There are spaces around operators, for example: `5 + 7` instead of `5+7`.
- Tabs have the same size as four spaces. Tabs are stored as tabs, not as spaces.
- Functions and static variables are in the same order in the .h and .cpp files.
- Try to make short functions, preferably no more than 50 lines.
- Do not define static variables in a function. Use class member variables instead (possibly static ones).
- Variables inside functions are declared at the point where they are needed, not all of them at the start of the function.

As a tester, you should be aware that these standards also give clues as to how code will fail under certain situations. For example, if machine-dependent ranges are hard-coded, you will see the resulting failure on one type of machine but not on another. Features that depend on values that could be machine-dependent should be tested on different machines.

In an embedded QA role, your responsibility is to check that the programmers have coding standards which they apply to their code. This is typically done by sampling files from the game code and doing a manual or automated check against the appropriate standards. This will not be possible, and may not be appropriate, if you are on a QA team removed from the developer, such as that of a publisher charged with testing milestone submissions from a third-party developer.

GAME QUALITY MEASUREMENTS

How good is "good" game software? Certainly, the number of defects in the code has something to do with goodness. The team's ability to find defects in its product is another factor to consider. A *sigma level* establishes the defectiveness of game code relative to its size, while the *phase containment* provides an indicator of how successful the team is at finding defects at their source, leaving fewer to escape to your players.

Six Sigma Software

A sigma level is one way to establish a goal for the outgoing quality of your game. For software, this measure is based on defects per million lines of code, excluding comments (also referred to as "non-commented source lines" or "NCSL"). The "lines of code" measure is often normalized to Assembly-equivalent lines of code (AELOC) in order to balance the different levels of abstraction across the variety of languages in use such as C, C++, Java, JavaScript, and Swift. The level of abstraction of each language is reflected in its multiplier. For example, each line of C code is typically regarded as the equivalent of three to four AELOC, whereas each line of Perl code is treated as about 15 AELOC. It is best to measure this factor based on your specific development environment and use that factor for any estimates or projections you need to make in the future. If you are using different languages for different parts of your game, multiply the lines of code for each portion by the corresponding language factor.

NOTE *Assembly code is the low-level instructions that are understood by the microprocessors running in your PC, game console, portable game device, or mobile phone. Assembly-equivalence refers to the number of Assembly language lines of code that are generated by compiling your game code in whatever language you wrote it in.*

Table 15.1 shows defect rates required to achieve a software quality measure anywhere between three and six sigma. Six sigma—only 3.6 defects per million lines of code—is typically regarded as an outstanding result, and getting in the 5.5 sigma range is very good. In case you think it is silly to worry about one million lines of code unless you are writing software for NASA, keep in mind that even mobile games can use a hundred thousand lines of code or more.

TABLE 15.1 Sigma table for various sizes of delivered software

Released Defects per (AELOC)				**Sigma Value**
20,000	**100,000**	**250,000**	**1,000,000**	
124	621	1552	6210	4.0
93	466	1165	4660	4.1
69	347	867	3470	4.2
51	256	640	2560	4.3
37	187	467	1870	4.4
27	135	337	1350	4.5
19	96	242	968	4.6
13	68	171	687	4.7
9	48	120	483	4.8
6	33	84	337	4.9
4	23	58	233	5.0
3	15	39	159	5.1
2	10	27	108	5.2
1	7	18	72	5.3
	4	12	48	5.4
	3	8	32	5.5
	2	5	21	5.6
	1	3	13	5.7
		2	9	5.8
		1	5	5.9
0	0	0	3	>6.0

Do not fool yourself by measuring your sigma on the sole basis of the open defects you know about in the product. This might reward poor testing which did not find many defects that still remaining in the game but would not reflect the experience your players will have. The defects being counted must include both the game defects you know about that have not been fixed, whatever defects your players have already found, and your projection of the number of defects that remain in the software that have not been discovered yet. It is best to wait anywhere from 6 to 18 months after shipping to calculate your sigma. If you still have a good result after that, continue to operate your projects in a similar manner by repeating what went "right" but also fix what went "wrong." If you have poor results, you should consider what changes you can make to avoid a repeat performance. You can start by going through the list of non-conformances that QA found during the project.

NOTE *You can find some sample game sigma tables in the spreadsheet file "SoftwareSigmaTable.xls" included with the book's companion files. which can be obtained by writing to info@merclearning.com.*

Phase Containment

Phase containment is the ability to detect faults in the project phase in which they were introduced. *Phase Containment Effectiveness* (PCE) is a measure of how well that is being done.

Faults that are found in the phase in which they are introduced are known as *in-phase faults*, or "errors." Faults that do not get caught in the same phase in which they are introduced are said to "escape" and become "defects." The principle is that if any subsequent work is derived from the faulty item, then a defect has occurred.

Errors are typically found by reviews, walkthroughs, or inspections. Defects are most noticeably found by testing and unhappy players, but they can also be found in reviews of downstream work products. For example, a code inspection issue might actually be the result of incorrect designs or requirements. Because the other work has already been done based on the fault, this is a defect.

PCE is typically tracked and reported by showing the faults found in each development phase. The faults are organized into columns for each phase in which they might be found. A coding fault cannot be detected in the requirements phase, because the code does not exist at that point. Calculate PCE by dividing the number of in-phase faults by the sum of faults found in all phases to come up with the PCE for that phase. From the data in Figure 15.1, the design phase PCE is calculated by dividing the number of faults found in the coding phase, 93, by the sum of all faults introduced by coding, which is 93 + 6 + 24 = 123. The result is 23/123 = 0.76. Figure 15.2 shows a graph summarizing the code PCEs for each phase.

	Phase where faults are found				
Phase created	REQMTS	DESIGN	CODING	TEST	PCE
REQMTS	114	27	4	15	0.71
DESIGN		93	6	24	0.76
CODING			213	105	0.67
Totals	114	120	223	144	

FIGURE 15.1: Game code phase containment data

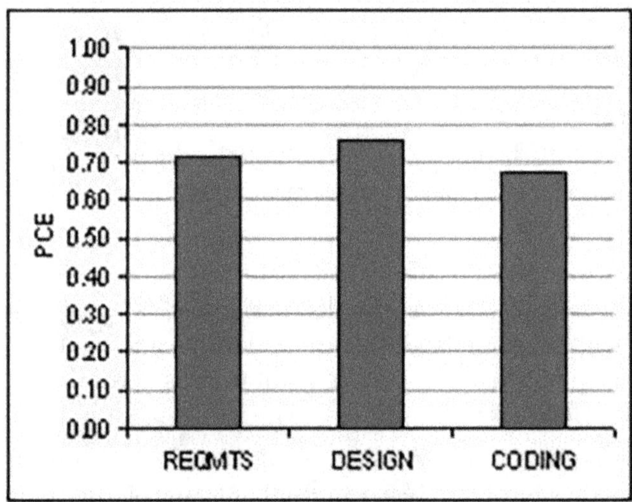

FIGURE 15.2 Game code phase containment graph

Alternatively, test results could be broken out into separate categories, as shown in Figure 15.3. These extra categories do not affect the PCE numbers or graphs, but this could be more convenient for data collection if different systems or categories are used for different release types. This data also helps the team to understand whether there will be additional testing activities that could further reduce the PCE numbers as more defects are found. In Figure 15.3, no Beta testing results are available to add to the table. The PCE numbers for requirements, design, and coding only represent the maximum possible value. New defects found in Beta testing will be sourced to these phases and reduce the corresponding PCEs.

Phase created	Phase where faults are found: REQMTS	DESIGN	CODING	TESTING: DEV TEST	DEMOS	ALPHA	BETA	PCE
REQMTS	114	27	4	11	3	1		0.71
DESIGN		93	6	19	5	0		0.76
CODING			213	90	10	5		0.67
Totals	114	120	223	120	18	6	0	

FIGURE 15.3 Game code phase containment date with expanded test categories

If this practice is useful for understanding how well the team is capturing defects in the game code, it should also be applied to the work produced by the testers. Figure 15.4 shows example PCE data for testing deliverables, and Figure 15.5 shows the corresponding graph.

Phase created	Phase where faults are found: DESIGN	SCRIPTING	CODING	EXECUTION	PCE
DESIGN	211	56	23	7	0.71
SCRIPTING		403	37	16	0.88
CODING			123	24	0.84
Totals	211	459	183	47	

FIGURE 15.4 Game test phase containment data

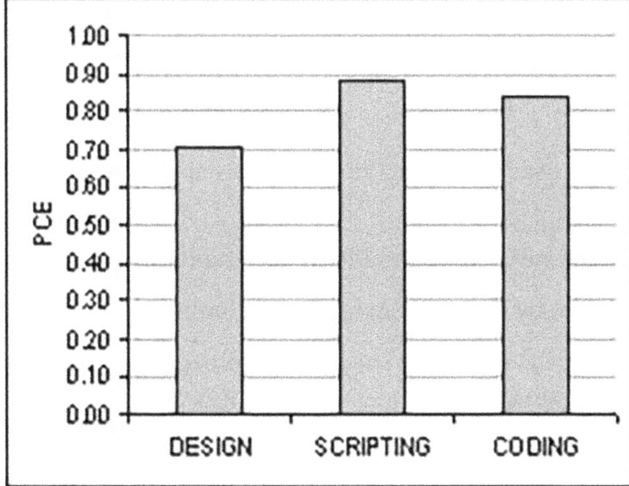

FIGURE 15.5 Game test phase containment graph

As the test PCE data shows, some faults in the tests do not get noticed until the test is executed on the game code. The problem might have been recognized as a test defect by the tester running the test, or it may have started out as a code defect before analysis and retesting uncovered the fact that the test was wrong, not the code. You can imagine how much more time consuming that is versus finding the defect before releasing the test.

Remember, this is not a measure of how well the executed tests perform. This is a measure of how well faults were captured in the test designs, scripts, or code. Any mistakes made in one of these activities will need to be repaired when they are eventually discovered. Test mistakes that do not get discovered could impact the quality of the game itself. A missing test, or a test that checks for the wrong result and passes, can send game bugs out to the paying public.

As with the sigma value, look for ways to improve your PCE. If you had 100% containment in all of your phases, you would only have to run each test once and they would all pass. Your players would not find any problems and you would never have to issue a patch. Think of the time and money that would save! Since the PCE is a function of both the faults produced and the faults themselves, you can attack a low PCE at both ends. Programmers can improve their ability to prevent the introduction of faults. Testers and QA can improve their ability to detect faults. In both cases, some basic strategies to address low PCE areas are as follows:

- Improve knowledge of the subject matter and provide relevant training.
- Have successful team members provide mentoring to less-successful members.
- Document methods used by successful individuals and deploy them throughout the team.
- Increase compliance with existing methods and standards.
- Add standards which, by design, help to prevent faults.
- Add checking tools that run during the creation process, such as color-coded and syntax-aware editors.
- Add checking tools that run after the creation process, such as stronger compliers and memory leak checkers.

QUALITY PLANS

Each game project should establish its own plan for how quality will be monitored and tracked during the project. This is typically documented in the SQAP. The SQAP contains no information about testing the game. That is covered in the game's Software Test Plan. The SQAP is strictly concerned with the independent monitoring and correction of product and process quality issues. It should address the following topics, most of which are covered in more detail below:

- QA personnel
- Standards to be used in the product
- Reviews and audits that will be conducted
- QA records and reports that will be generated
- QA problem reporting and corrective actions
- QA tools, techniques, and methods
- QA metrics
- Supplier control
- QA record collection, maintenance, and retention
- QA training required
- QA risk management

QA Personnel

Begin this section by describing the organizational structure of the QA team. Show who the front-line QA engineers work for and whom the head of QA reports to. Identify at which level the QA reporting chain is independent from the person in charge of the game development staff. This helps to establish a path for escalating QA issues and identifies which key relationships should be nurtured and maintained during the project. A good rapport between the QA manager and the development director will have a positive effect on both the QA staff and the development staff.

Describe the primary role of each person on the QA team for this project. List what kinds of activities each of them will be involved in. Be as specific as possible. If a person is going to be responsible for auditing the user interface screens against the company's UI standards, then say that. If another person is going to take samples of code and check them with a static code analysis tool, then say that. Use a list or a table to record this information.

Strictly speaking, QA and testing are separate, distinct functions. QA is more concerned with auditing, tracking, and reporting, whereas testing is about the development and execution of tests in the relentless pursuit of finding operational defects in the game. However, depending on the size and skill of your game project team, you may not have separate QA and test teams. It is still best to keep those two plans separate even if some or all of the same people are involved in both kinds of work.

Standards

Two types of standards should be addressed in this section: product standards and process standards. Product standards apply to the *things that are produced* as part of the game project. This includes code, graphics, printed materials, and platform manufacturer guidelines. Process standards apply to the *way things are produced* as part of the game project. This includes file naming conventions, code formatting standards, and maintenance of evolving project documents such as the TDD. Document all of the standards that apply, as well as which items they apply to. Then describe how the QA staff will monitor them and follow up on any discrepancies.

Reviews and Audits

The kinds of reviews performed by QA are not the same as those that developers or testers would do for code or test designs. A QA review is usually done by a single QA engineer who evaluates a work product or ongoing process against some kind of reference, such as a checklist or standard. QA reviews and audits span all phases and groups within the game project.

Project documents, project plans, code, tests, test results, designs, and user documentation are all candidates for QA review. QA should also audit work procedures used by the team. These can include the code inspection process, file backup procedures, and the use of tools to measure game performance over a network.

Reviews and audits can be performed on the results of the process, such as checking that all required fields in a form are filled in with the right type of data and that required signatures have been obtained. Another way to audit is to observe the process in action. This is a good way to audit peer reviews, testing procedures, and weekly backups. Procedures that occur very infrequently, such as restoring project files from backup, can be initiated by QA to make sure that the capability is available when it is needed.

QA itself should be subject to independent review (remember Rule Two). If you have multiple game projects going on, each project's QA team can review the work of the other to provide feedback and suggestions to ensure that they are doing what they documented in the SQAP. If no other QA team exists, you can have someone from another function such as testing, art, or development use a checklist to review your QA work.

The QA activities identified in this section of the SQAP should be placed on a schedule to ensure that the QA people will have the time to do all of the activities they are signed up for. These activities should also be coordinated with the overall project schedule and milestones so you can count on the work products or activities that are being audited to be available at the time you are planning to audit them.

To be considerate, planned QA activities that will disrupt other peoples' work, such as restoring backups or sitting down with someone to review a month's worth of TDD updates, should be incorporated into the overall project schedule so the people affected will be able to set aside the appropriate amount of time for preparing and participating in the audit or review. This is not necessary for activities such as sitting in on a code review, because the code review was going to take place whether or not you were there.

Feedback and Reports

The SQAP should document what kinds of reports will be generated by software quality assurance (SQA) activities and how they will be communicated. Reporting should also include the progress and status of SQA activities against the plan. These get recorded in the SQAP along with how frequently the QA team's results will be reported and in what fashion. Items that require frequent attention should be reported on regularly. Infrequent audits and reviews can be summarized at longer intervals. For example, the QA team might produce weekly reports on test result audits, but produce quarterly reports on backup and restoration procedure audits. Test result audits would begin shortly after testing starts and continue through the remainder of the project. Backup and restoration audits could start earlier, once development begins.

SQA reporting can be formal or informal. Some reports can be sent to the team via email, while others may aggregate into quarterly results for presentation to company management at a quarterly project quality review meeting.

Problem Reporting and Corrective Action

SQA is not simply done for the satisfaction of the QA engineers. The point of SQA is to provide a feedback loop to the project team so that they are more conscientious about the importance of doing things the right way. This includes keeping important records and documents complete and up to date. It is up to QA to guide the team or the game company in determining which procedures and work products benefit the most from this compliance. Once an SQA activity finds something to be non-compliant, a problem report is generated.

Problem reports can be very similar to the bug reports you write when a tester finds a defect in the software. They should identify which organization or individual will be responsible and describe a time frame for resolving the issue. The SQAP should define what data and statistics on non-compliant issues should be reported on, as well as how and when they are to be reviewed with the project team.

History has shown, unfortunately, that some project members might be more reluctant to spend time closing SQA problems because they have their “real job” to do, such as development, testing, and art creation. As a consequence, it is a good idea to define the criteria and process for escalating unresolved issues. Similarly, there should be a defined way for resolving issues with products that cannot be fixed within the game team, such as software tools or third-party APIs.

In addition to addressing compliance issues one at a time, SQA should also look for the causes of negative trends or patterns and suggest ways to reverse them. This includes process issues, such as schedule slippages, and product issues, such as game asset memory requirements going over budget. The SQAP should document how the QA team will detect and treat the causes of such problems.

Tools, Techniques, and Methods

Just like development and testing, the QA team can benefit from tools. Since QA project planning and tracking needs to be coordinated with the rest of the project, it is best if they use the same project management tools as the rest of the game team. Likewise, tracking issues found in QA audits and reviews should be done under the same system used for code and test defects. Different templates or schemas might be needed for QA issue entry and processing, but this will keep the team software licensing and operation costs down and make it easy for the rest of the team to access and update QA issues.

Some statistical methods might be useful for QA analysis of project and process results. Many of these methods are supported by tools. Such tools and methods should be identified in the SQAP. For example, Pareto charts graph a list of results in descending order. The bars furthest to the left are the most frequently occurring items. These are the issues you should spend your time on first. If you are successful at fixing them, the numbers will go down and other issues will replace them to the left of the chart. You can go on forever addressing the issue at the left of the chart because there will always be one. This is kind of like trying to clean your room. At some point in time, you can decide the results are "good enough" and move on to some entirely different result to improve.

Figure 15.6 shows an example Pareto chart of the number of defects found per thousand lines of code (KLOC) in each major game subsystem. The purpose of such a chart could be to identify which portion of the code would benefit the most from using a new automated checking tool. Because there are costs associated with new technologies—purchasing, training, extra effort to use the tool, and so on—it should be introduced where it would have the greatest impact. In this case, start with the rendering code.

Another useful software QA method is to plot control charts of product or process results. The control chart shows the average result to expect and "control limit" boundary lines for the set of data provided. Any items outside of the control limits fall beyond the range of values that would indicate they came from the same process as the rest of the data. This is like having a machine that stamps metal squares a certain way, but every once in a while, one comes out very different from the others. If you have the right amount of curiosity to be a QA person, you would want to know why the square comes out wrong some of the time. The same is true for anomalies in software The control chart reveals results that should be investigated to understand their cause. It might simply be a result of someone entering the wrong data (such as date, time, size, and defects). Figure 15.7 shows an example control chart for new lines of delta (added or deleted) code changes in the game each week. The numbers are in KLOC.

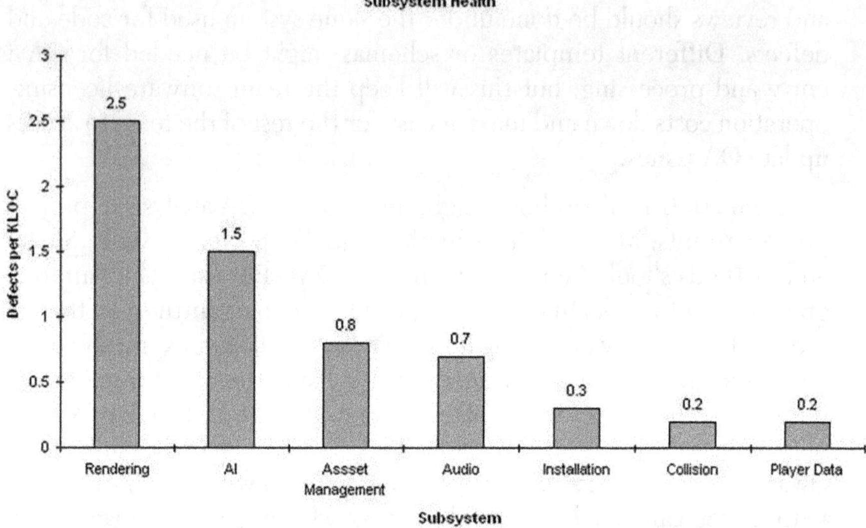

FIGURE 15.6 Pareto chart of defects per KLOC for each game subsystem

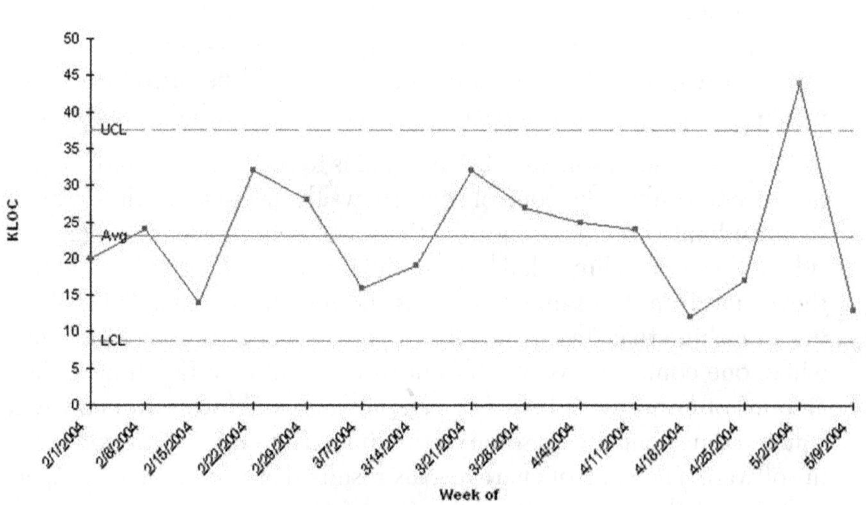

FIGURE 15.7 Control chart of weekly code change in KLOC

The solid line running across the middle of the chart is the average value for the data set. The two dashed lines labeled UCL and LCL represent the Upper Control Limit and the Lower Control Limit, respectively. These values are calculated from the date set as well. The data point for the week of "5/2/2004" lies above the UCL. This is a point that should be investigated.

"Thanks, Bud!"

On one project, there was a noticeable dip in the number of defects submitted in one particular week. This was a good result for the developers, but bad for the testers. A quick investigation revealed that "Bud"—an especially productive tester—had been on vacation that week. The test data for the rest of the team was within the normal range. Legitimately bad results should be understood and subsequently prevented from happening in the future. However, especially good results are just as important to understand so that they can be imitated. Additional tools and techniques can be identified in the SQAP for these purposes. This result also suggests that the data could be reported in a different way, such as defects per tester, to account for inevitable fluctuations in staffing. This could replace the original chart or be used in addition to it.

Supplier Control

Your game is not just software: it is a customer experience. The advertisements in the store, the game packaging, the user manual, and the game's website are all part of that experience. In many cases, these items come from sources outside of the core development team. These are some of your "suppliers." Their work is subject to the same kinds of mistakes you are capable of producing on your own. You may also have software or game assets supplied to you that you use within the game, such as game engines, middleware, art, and audio files.

In both of the cases above, QA should play a role in determining that the supplied items are "fit for use." This can be done in the same way internal deliverables are evaluated. The QA team can also evaluate the supplier's capability to deliver a quality product by conducting on-site visits to evaluate the supplier's processes. When you go to the deli, the food is laid out nicely in the display case. You also appreciate the fact that a food inspector has evaluated the plant where the food originates from to see that it is uncontaminated, and that it is produced in a clean and healthy environment. The same should be true for game-related software and materials that are supplied to you from other companies.

Training

If new tools, techniques, or equipment are going to be used in the development of the project, it may be necessary for one or more QA personnel to become acquainted so they can properly audit the affected deliverables and activities. The impact of the new technologies may affect QA preparation as well, such as requiring new audit checklists to be created or new record types to be defined in the audit entry and reporting system.

The QA training should be planned and delivered in time for QA to conduct any activities related to work products or processes using the new technology. If the team is already having an in-house course delivered, then add some seats for QA. If the team is inventing something internally, try to get a briefing from one of the inventors. Some tools and development environments come with their own tutorials, so get some QA licenses and allocate time to go through the tutorial.

New tools or techniques identified for QA-specific functions should be accompanied with appropriate training. Identify these, document them in the SQAP, and get your training funded.

Risk Management

Risk management is a science all unto itself. In addition to all of the risks involved with developing your game, there are also risks that could hamper your team's QA efforts. Some typical SQA risks are as follows:

- project deliverables go out of sync with planned audits
- QA personnel diverted to other activities, such as testing
- lack of an independent QA reporting structure
- lack of organization commitment to take corrective actions or to close out issues raised by QA
- insufficient funding for new QA technologies
- insufficient funding for training in new development or QA technologies

It is not enough to list your risks in the SQAP. You also need to identify the potential impact of each risk and any action plans you can conceive to describe how you would proceed if the risk occurs or persists.

Software quality is certainly affected by testing, but there are other activities that can impact quality sooner and less expensively. Various forms of peer reviews can find faults before they escape to other phases of the project. Standards can be defined and enforced as a way to prevent defects from being introduced into the game, many of which are difficult to detect by testing. Measures such as sigma value and phase containment provide a foundation on which you can build your improvement goals. The Software Quality Assurance organization carries out activities according to a plan that monitors and promotes the use of these techniques and measures. Their cost must be weighed against the consequences and costs of releasing a poor-quality game.

For more information and resources on software quality, explore the American Society for Quality at www.asq.org.

EXERCISES

1. Your game code size is 200,000 AELOC. It had 35 defects you knew about when you released it. The people who bought it have reported 17 more. At what sigma level is your game code?
2. Describe the differences between the Leader's role in a walkthrough and the Moderator's role in a Fagan Inspection.
3. Add the following defects found in Beta testing to the data in Figure 15.3: Requirements—5, Design—4, Coding—3. What are the updated code PCEs for the requirements, design, and coding phases?
4. What are the responsibilities of a tester in a peer review and how can they contribute to the review process?

REFERENCES

Caminos, Rob. 2004. "Cross-Platform User Interface Development." *Game Developer*, March 25, 2004. *https://www.gamedeveloper.com/design/cross-platform-user-interface-development*.

Crosby, Philip B. 1980. *Quality is Free: The Art of Making Quality Certain*. United States: New American Library.

Fagan, M.E. 1986. "Advances in Software Inspections." *IEEE Transactions on Software Engineering* vol. SE-12, no. 7, pp. 744-751. *https://doi.org/10.1109/TSE.1986.6312976*.

Van Dongen, Joost. 2017. "The Ronimo Coding Style Guide." *Game Developer*, July 11, 2017. *https://www.gamedeveloper.com/programming/the-ronimo-coding-style-guide*.

Appendix A

Odd-Numbered Answers to Exercises

Chapter 1 — Your Role on the Game Development Team

1. There will be more testers toward the end of the game project.
3. GDDs are vast and updated very frequently. Printing one out would be simply a waste of paper.
5. A patch usually contains only bug fixes, while an update may include both bug fixes and new game content.
7. Alpha, Beta, and Gold. They should be well defined so that the developer and publisher have very clear criteria against which to test each milestone candidate.

Chapter 2 — The Basics of Game Testing

1. (a) J, (b) P, (c) J, (d), P
3. False
5. The test suite is a list of individual test cases designed to test whether specific features and functions of the game are working as designed. If a build fails a test, then it is not working as designed, and the tester should write a bug report about that test failure.

Chapter 3 — Bug Report Writing and Defect Tracking

1. The expected result is what the designer intended or what the player would reasonably expect when performing a certain action in the game. The actual result is what happened in the game that deviated from such expectations, so the actual result describes the bug.
3. False
5. False

Chapter 4 — How Bugs Happen

1. Game testing is important because games get made wrong.
3. Correct answers to this question should include the following ideas: when you are placed in a particular location in the game world, when you type in a name for something in the game (such as a player, town, and pet), when you change a game option (such as the language and difficulty), when you gain a new ability (such as a skill, level, job, and unlocked item), or when you set the selling price of an item.
5. `RespawnItem` defect opportunities:
   Function — 1 through 15 (random selection), 16-20 (setup and use flags), 21-22 (play respawn sound)

   Assignment — 9, 10, 13, 16, 23

   Checking — 2, 6, 10, 13

   Timing — 22

   Build/Package/Merge — 17

   Algorithm — 12, 28, 19

   Documentation — 7 (a literal string is used to report an error)

   Interface — 0, 7, 20, 22

Chapter 5 — The Phases of Game Quality Assurance

1. The main responsibilities of a lead tester are managing the test team, designing and implementing the overall project test plan, and "owning" the bug database.

3. False
5. A test plan defines the overall structure of the testing cycle. A test case is one specific question or condition the code is operated and evaluated against.

Chapter 6 — Exploratory Testing and Gameplay Testing

1. False
3. Free testing is an unstructured search for software defects. It results in additional bugs being discovered. Gameplay testing is a structured attempt to judge the quality, balance, and fun of a game. It results in suggestions and feedback that the designers can use to tweak and polish the game.
5. This is gameplay testing. The testers are playing the game, not testing the game.

Chapter 7 — The Two Rules of Test Management

1. A great deal of money can be at stake when you develop a game.
3. Check your work against a pre-written checklist, and have other testers check your work (and you can check theirs).
5. Low morale means low productivity, low engagement, and high turnover.

Chapter 8 — Combinatorial Testing

1. Full combinatorial tables provide all possible combinations of a set of values with each other. The size of such a table is calculated by multiplying the number of choices being considered (tested) for each parameter. A pairwise combinatorial table does not have to incorporate all combinations of every value with all other values. It is "complete" in the sense that somewhere in the table there will be at least once instance of any value being paired up in the same row with any other value. Pairwise tables are typically much smaller than full combinatorial tables; sometimes hundreds or thousands of times smaller.

3. Use the template for three parameters with three values and four parameters with two values in Appendix D to arrive at Table A.1. The cells with "*" can have either a "Yes" or "No" value and your table will still be a correct pairwise combinatorial table.

TABLE A.1 *FIFA 15* match settings test table with seven parameters

Test	Half Length	Referee	Weather	Difficulty	Pitch Wear	Game Speed	Offsides
1	4 min	Lenient	Dry	Amateur	None	Slow	**On**
2	10 min	Average	Rainy	Legendary	High	Slow	**On**
3	20 min	Strict	Snowy	Amateur	High	Fast	**On**
4	4 min	Average	Snowy	Legendary	None	Fast	**Off**
5	10 min	Strict	Dry	Legendary	None	Fast	*
6	20 min	Lenient	Rainy	Legendary	None	Fast	*
7	4 min	Strict	Rainy	Amateur	High	Slow	**Off**
8	10 min	Lenient	Snowy	Amateur	High	Slow	**Off**
9	20 min	Average	Dry	Amateur	High	Slow	**Off**

5. If you provided the right parameters and values to Allpairs, you should get the tests shown in Table A.2 (the "pairings" column has been left out). If your input table had the parameters in a different order that was used for this solution, verify that you have the same number of test cases as Table A.2. A total of 540 full combinations have been reduced to 23 pairwise tests. If your result does not seem right, redo the input table following the same ordering of the parameters that appear in Exercise 3 and try again.

TABLE A.2 Kingturn RPG game options settings

Case	Sound	Difficulty	Perma Knockout	Pinch Zoom
1	On	Casual	On	Slowest
2	Off	Casual	Off	Slower
3	Off	Normal	On	Slowest
4	On	Normal	Off	Slower
5	On	Strategist	On	Default
6	Off	Strategist	Off	Faster
7	Off	Master	Off	Default
8	On	Master	On	Faster
9	On	King	Off	Fastest
10	Off	King	On	Slowest
11	Off	Casual	On	Fastest
12	~On	Casual	Off	Slowest
13	~On	Normal	On	Slower
14	~Off	Normal	~Off	Default
15	~Off	Strategist	~On	Slower
16	~On	Strategist	~Off	Fastest
17	~Off	Master	~On	Fastest
18	~On	Master	~Off	Slowest
19	~On	King	~On	Default
20	~Off	King	~Off	Faster
21	~On	Casual	~On	Faster
22	~Off	Casual	~Off	Default
23	~On	Normal	~Off	Faster
24	~Off	Normal	~On	Fastest
25	~Off	Strategist	~Off	Slowest
26	~On	Master	~On	Slower
27	~Off	King	~Off	Slower

Chapter 9 – Test Flow Diagrams

1. Your answer should at least describe the following kinds of changes:

l. Change "Ammo" to "Arrows" and "Gun" to "Bow."

m. "DropSound" would be different for the arrows (rattling wood sound) than for the bow (light "thud" on grass, "clank" on cobblestone), so need two distinct events for "DropArrowSound" and "DropBowSound."

n. If you have both the bow and some arrows, dropping the bow will not cause you to lose your arrows, so flow 8 should connect to the "HaveAmmo" state.

o. It is not really possible to pick up a loaded bow, so eliminate the "GetLoadedGun" flow (9).

p. "ShootGun" (now "ShootBow") may make more of a "twang" or "whoosh" sound if there is no arrow, so change "ClickSound" to "NoArrowSound" or something similarly descriptive.

q. Firing a bow requires more steps than shooting a gun. You could add some or all of the states and flows for the steps of taking an arrow from the quiver, loading the arrow on the bowstring, pulling the string, aiming, and releasing the arrow. Your reason for doing this should remain consistent with the purpose of the TFD. For example, with a bow and arrows, you could load the arrow to go to an "ArrowLoaded" state, but then unload the arrow to go back to "HaveBowHaveArrows" to make sure the arrow you did not fire was not deducted from your arrow count.

3. From Exercise 2, your updated TFD should at least have a "GetWrongAmmo" flow going from "HaveGun" to a new "HaveGunWrongAmmo" state. From that state you would have a "DropWrongAmmo" flow going back to "HaveGun" and a "ShootGun" flow with a "ClickSound" action looping back to "HaveGunWrongAmmo" the same way flow 3 does with the "HaveGun" state. Your minimum path must include all of the new flows, passing through the "HaveGunWrongAmmo" state. For the baseline path generation, you may choose the same baseline that applies to Figure 9.10 or define a different one. At some point, you need to have a

derived path that gets to the "HaveGunWrongAmmo" state and passes through the "ShootGun" loop. Swap your test case with a friend and check each other's results step by step. It may help to read out loud as you go along and trace the flows that are covered with a highlighter.

Chapter 10 — Cleanroom Testing

1. It is possible to have the same exact test case appear more than once in a Cleanroom test set. This would typically involve values that have high usage frequencies but, like the lottery, it is also possible that infrequent value combinations will be repeated in your Cleanroom table.
3. Your answers will depend upon the game you choose.
5. From Exercise 4, you should have produced inverted usage values for the Casual user profile as follows:
   Look Sensitivity: 1 — 32%, 3 — 4%, 10 — 64%

   Look Inversion: Inverted — 90%, Not Inverted — 10%

   AutoLook Centering; Enabled — 70%, Disabled — 30%

   Crouch Behavior: Hold — 20%, Toggle — 80%

   Clench Protection: Enabled — 75%, Disabled — 25%

   The random number set that was used to produce the table in Figure 10.12 generates the following inverted usage test data:

   a. Look Sensitivity = 1, Look Inversion = Inverted, AutoLook Centering = Enabled, Crouch Behavior = Toggle, Clench Protection = Enabled

   b. Look Sensitivity = 10, Look Inversion = Inverted, AutoLook Centering = Enabled, Crouch Behavior = Toggle, Clench Protection = Enabled

   c. Look Sensitivity = 1, Look Inversion = Inverted, AutoLook Centering = Enabled, Crouch Behavior = Toggle, Clench Protection = Enabled

   d. Look Sensitivity = 10, Look Inversion = Inverted, AutoLook Centering = Enabled, Crouch Behavior = Toggle, Clench Protection = Enabled

   e. Look Sensitivity = 3, Look Inversion = Inverted, AutoLook Centering = Disabled, Crouch Behavior = Toggle, Clench Protection = Disabled

f. Look Sensitivity = 10, Look Inversion = Inverted, AutoLook Centering = Enabled, Crouch Behavior = Toggle, Clench Protection = Enabled

7. The path produced from the inverted usage values will depend on the random numbers you generate. Ask a friend or a classmate to check your path and offer to check theirs in return.

Chapter 11 – Test Trees

1. The bug fix affects "sound," "Orks," and "weapon," so you should run the collection of tests associated with the following nodes on the tree: Options — Sound

   Game Modes — Skirmish — Races (Orks)

   Races — (Orks)

3. You should indicate on your diagram how many spells are required to enable each new lesson and notice where the same spell unlocks different lessons. A correct tree is drawn with a vertical orientation in Figure A.1

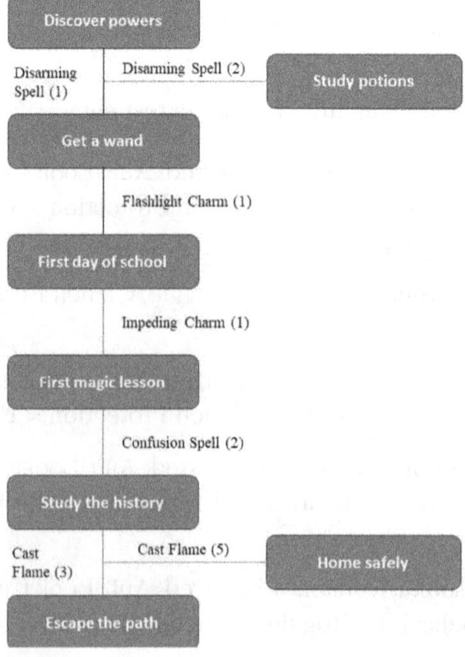

FIGURE A.1 *School of Wizardry* test tree solution

Chapter 12 — Defect Triggers

1. The answer is specific to the reader.
3. Representing the "snap back" behavior on the TFD requires a state to represent your avatar at the starting location and another state to represent your avatar standing at a gun or ammo location. A "MoveToGun" flow would take you from the "PreMatch" location to the "standing" location. A flow with a "PrematchTimerExpires" event would take you from your standing location to the "NoGunNoAmmo" state, accompanies by an action describing the "snap back" to the starting position. For the case where you do not move from the initial spawning location, add a "PrematchTimer Expires" flow from the "PreMatch" location to "NoGunNoAmmo" but without the snap back action.
5. Besides the "Normal" trigger testing, which you are accustomed to, here are some ways to utilize other defect triggers for this hypothetical poker game:

   Startup: Try to do things during the intro and title screens, try to bet all of your chips on the very first hand, and try to play without going through the in-game tutorial.

   Configuration: Set the number of players at the table to the minimum or maximum, set the betting limits to the minimum or maximum, play at each of the difficulty settings available, and play under different tournament configurations.

   Restart: Quit the game in the middle of a hand and see if you have your original chip total when you re-enter the game, create a split pot situation where one player has wagered all of their chips but other players continue to raise their bets, and save your game and then reload it after losing all of your money.

   Stress: Play a hand where all players bet all of their money, play for long periods of time to win ridiculous amounts of cash, take a really long time to place your bet or place it as quickly as possible, and enter a long player name or an empty one (0 characters).

Exception: Try to bet more money than you have, try to raise a bet by more than the house limit, and try using non-alphanumeric characters in your screen name.

Chapter 13 – Regression Testing and Test Reuse

1. The first stage of your defect model should have states for an undamaged meatshield, damaged meatshield, and a destroyed (by the chainsaw) meatshield, as shown in Figure A.2.

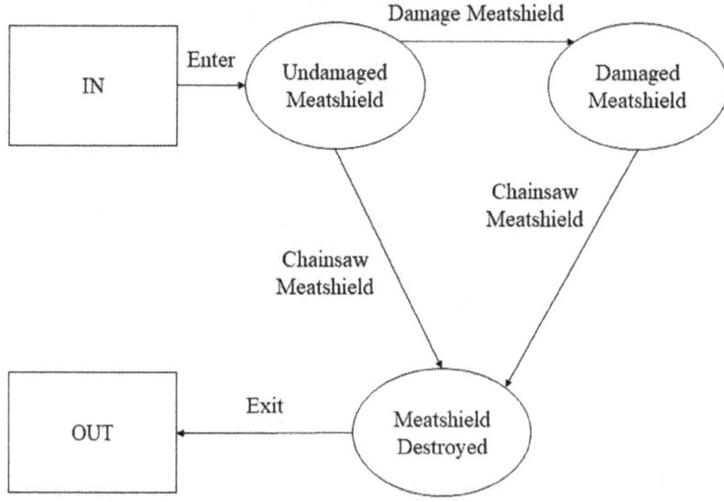

FIGURE A.2 Basic meatshield defect model TFD

You should also tack on a few more states and flows to make sure you check what happens if the meatshield is dropped and picked up, and to verify that the meatshield can actually be destroyed. Figure A.3 shows a TFD that incorporates these added elements. For the final touch, put actions where they are appropriate.

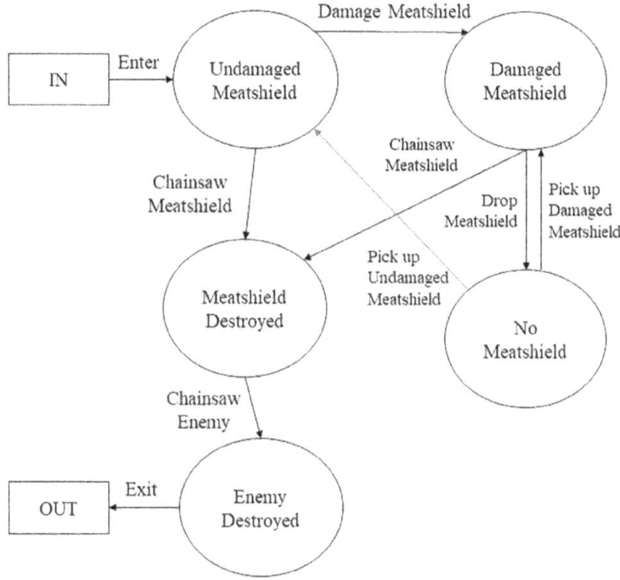

FIGURE A.3 Enhanced meatshield defect model TFD

3. The following events or situations are missing the player's name. Either blank information is shown or only the player's jersey number (8) is shown:

   a. At 0:59 after the captains meet the referee at midfield, the Bohemians' FC lineup is shown. Player 8's number is shown without a name.

   b. At 1:03, a graphic shows the Bohemians' FC formation, but there is no name below the number 8 jersey.

   c. At 1:32, scrolling through the team roster shows the number 8 player is wearing the captain's armband, but his name does not appear.

   d. As the game begins, each player's name appears above them when they have possession of the ball. From 1:50 to 1:55, player 8 dribbles the ball towards the goal, but his name does not appear. This also occurs when he attempts a shot from 1:58 to 2:00.

   e. At 2:35, player 8 gets the ball again and is subsequently fouled. A free kick is awarded and when scrolling through players to change the kicker, player 8's name is blank.

f. When player 8 takes the kick from 2:55 to 3:00, his name never appears.

g. At 3:20, player 8 receives the ball at midfield and heads towards the goal, scoring at 3:28. A pop-up badge appears on the screen to indicate the goal scorer, but only the number 8 is shown.

h. Once play resumes at 3:42, the goal notification pops up above the scoreboard. The time of the goal is properly shown, but the name of the goal scorer is not. You can compare that to what's displayed at 4:30 when the opposing team scores.

i. The video skips ahead to the end of the match and again, player 8's name is missing from his goal.

j. After the match, at 5:28, the Player Ratings screen shows the players' names below their image, except for player 8 (above and to the right of the goalkeeper).

Chapter 14 — Testing by the Numbers

1. The original two testers, B and Z, were progressing at a steady rate that was not quite enough to keep up with their goal. Tester D was added in January, but the team's total output was not improved. This could be due to effort diverted from testing to provide support to D or to verify that D's tests were done properly. On January 8, C and K were added to the test team, while B took a day off. We can presume C and K knew what they were doing, as the group output went up, and they almost met the goal. K and Z did not participate after that, and the output when back down, even as B returned. Ultimately, only D was left on the project, as presumably the others were reassigned to more vital testing. D completed seven tests on the 12th, but it remains to be seen if he can sustain this level of output until the project can get its testing staff back up to where it should be. The two important observations here are that you cannot treat every tester as an identical plug-in replacement for any other tester (they each have their own strengths and skill sets) and adding more testers does not guarantee a proportional increase in team output, especially during the first few days.

3. Tester C made the best use of their test opportunity to find the most defects per test. However, other testers such as B and Z were able to perform many more tests and find a few more defects. Since "Best Tester" is based on the combined overall contribution to tests completed

and defects found, C is not in the running. It is still important to identify C's achievements and to recognize them. If B and Z could have been as "effective" as C, they could have found about six more defects each—a very significant amount.

5. Some positive aspects of measuring participation and effectiveness are as follows:
   - Some people will do better if they know they are being "watched."
   - Some people will use their own data as motivation to improve on their numbers during the course of the project.
   - It provides a measurable basis for selecting "elite" testers for promotion or special projects (as opposed to favoritism, for example).
   - Testers seeking better numbers may interact more with developers to find out where to look for defects.

Some negative aspects are as follows:

- Effort is required to collect and report the tester data.
- It can be used as a tool to punish certain testers.
- It may unjustly lower the perceived "value" of testers who make important contributions in other ways, such as mentoring.
- It could lead to jealousy if one person constantly wins.
- Testers may argue over who gets credit for certain defects (which hinders collaboration and cooperation).
- Some testers will figure out a way to exceed their individual numbers (such as choosing easy tests to run) without really improving the overall test capabilities of the team.

Chapter 15 — Software Quality

1. Your total released defects are 35 + 17 = 52. The table in Figure 15.1 has a column for 100,000 but not for 200,000, so double the defect count values in the 100,000 column. A defect count of 66 indicates a 4.9 sigma level and 48 is 5 sigma. Your 52 defects do not reach the 5 sigma level, so your game code is at 4.9 sigma.

3. The new PCE for the requirements phase is 0.69. The new PCE for design is 0.73. The new code PCE is 0.66.

Appendix B

Sample Test Suite: RTS Building Checklist

The following is a portion of test suite written to test all the building functionality for one faction of a 3D real-time strategy game. It attempts to isolate each graphic and audio asset associated with each building, as well as the individual functions of each building. Note that each question is written such that a "yes" answer means a pass condition to the test case, and a "no" answer means a fail—and a possible defect.

NOTE *The full test suite is included with the book's companion files, which can be obtained by writing to info@merclearning.com.*

Dragon Building Checklist			
Name ______________________ **Build** ________	**pass**	**fail**	**comments**
(**NOTE:** Start all buildings with a single peasant.)			
Peasant Hut			
Select a Peasant. Choose "Peasant Hut."			
Building cost OK?			
Text OK?			
Can't build if price not met?			

Building footprint graphic OK?			
Can't place building if footprint is red?			
Can rotate building 360 degrees?			
Right-click places building?			
Building site appears in Peasant's LOS?			
Peasant constructs building?			
Construction animation OK?			
Phase one graphic OK?			
Phase two graphic OK?			
Phase three graphic OK? (if present)			
Completed building graphic OK?			
"Building Finished" audio OK?			
Select building.			
Toolbar graphics OK?			
Text OK?			
Can destroy building?			
Can cancel building destruction?			
Confirming destroys building?			
Animation OK?			
Audio OK?			
Build new building of the same type.			
Place it away from first building.			
Alt+right click makes peasant run to building site?			
Select new building. Right-click on ground nearby to set rally point.			
Flag appears?			
Spawning units walk to rally point?			

Select building again. Alt+right-click to set new rally point further away.			
Flag appears?			
Spawning units run to rally point?			
Select building again. Set rally point on rice field.			
Spawning units gather rice?			
Select building again. Set rally point on water source.			
Spawning units gather water?			
Create second Peasant Hut. Select first Peasant Hut.			
Peasant toggle icon OK?			
Text OK?			
Stops peasants from spawning from this building?			
Select building again. Right-click on any barracks.			
Spawning units enter barracks for training?			
Have fire-damaging enemies attack building. (Serpent Raiders are good for this.)			
Building burns?			
Animation OK?			
Audio OK?			
Building is destroyed in stages?			
Stage one graphics OK?			
Audio OK?			
Stage two graphics OK?			
Audio OK?			
Stage three graphics OK (if present)?			
Audio OK?			
Building leaves rubble when destroyed?			
Rubble disappears after a while?			

Well			
Select a Peasant. Choose "Well."			
Building cost OK?			
Text OK?			
Can't build if price not met?			
Can't build if prerequisites not met?			
Building footprint graphic OK?			
Can't place building if footprint is red?			
Can rotate building 360 degrees?			
Right-click places building?			
Building site appears in Peasant's LOS?			
Peasant constructs building?			
Construction animation OK?			
Phase one graphic OK?			
Phase two graphic OK?			
Phase three graphic OK? (if present)			
Completed building graphic OK?			
"Building Finished" audio OK?			
Select building.			
Toolbar graphics OK?			
Text OK?			
Can destroy building?			
Can cancel building destruction?			
Confirming destroys building?			
Animation OK?			
Audio OK?			
Select a Peasant. Click "Well."			
Peasant gathers water?			

Have fire-damaging enemies attack building. (Serpent Raiders are good for this.)			
Building burns?			
Animation OK?			
Audio OK?			
Building is destroyed in stages?			
Stage one graphics OK?			
Audio OK?			
Stage two graphics OK?			
Audio OK?			
Stage three graphics OK (if present)?			
Audio OK?			
Building leaves rubble when destroyed?			
Rubble disappears after a while?			
Dojo			
Select a Peasant. Choose "Dojo."			
Building cost OK?			
Text OK?			
Can't build if price not met?			
Can't build if prerequisites not met?			
Building footprint graphic OK?			
Can't place building if footprint is red?			
Can rotate building 360 degrees?			
Right-click places building?			
Building site appears in Peasant's LOS?			
Peasant constructs building?			
Construction animation OK?			
Phase one graphic OK?			
Phase two graphic OK?			

Phase three graphic OK? (if present)			
Completed building graphic OK?			
"Building Finished" audio OK?			
Select building.			
Toolbar graphics OK?			
Text OK?			
Can destroy building?			
Can cancel building destruction?			
Confirming destroys building?			
Animation OK?			
Audio OK?			
Build new building of the same type.			
Place it away from first building.			
Alt+right-click makes peasant run to building site?			
Select new building. Right-click on ground nearby to set rally point.			
Flag appears?			
Send unit in for training.			
Exiting units walk to rally point?			
Select building again. Alt+right-click to set new rally point further away.			
Flag appears?			
Send unit in for training.			
Exiting units run to rally point?			
Select building again. Right-click on any other barracks.			
Flag appears?			
Send unit into Dojo for training.			
Exiting units enter second barracks for training?			
Select a peasant, then hover cursor over building.			
“Train ________” hotspot is entire building footprint?			

Send unit into building.			
Cancel button graphics OK?			
Text OK?			
Unit exits building when you click "Cancel?"			
Resource cost recovered?			
Have fire-damaging enemies attack building. (Serpent Raiders are good for this.)			
Building burns?			
Animation OK?			
Audio OK?			
Building is destroyed in stages?			
Stage one graphics OK?			
Audio OK?			
Stage two graphics OK?			
Audio OK?			
Stage three graphics OK (if present)?			
Audio OK?			
Building leaves rubble when destroyed?			
Rubble disappears after a while?			
Build four of this type of building.			
Send a peasant in to train in the first building; buy one each of the building's three techniques in the other three buildings. (Do this as close to simultaneously as you can. Use the "yinyang" cheat to get enough yin or yang points.)			
Peasant enters building?			
Peasant progress meter OK?			
Technique 1 progress meter OK?			
Technique 2 progress meter OK?			
Technique 3 progress meter OK?			
First-level unit exits building?			

Exit game. Create new game.			
Build four of this type of building.			
Send a first-level unit in to train in the first building; then buy one each of the building's three techniques in the other three buildings. (Do this as close to simultaneously as you can. Use the "yinyang" cheat to get enough yin or yang points.)			
First-level unit enters building?			
First-level unit progress meter OK?			
Technique 1 progress meter OK?			
Technique 2 progress meter OK?			
Technique 3 progress meter OK?			
Second-level unit exits building?			
Exit game. Create new game.			
Build four of this type of building.			
Send a second-level unit in to train in the first building; then buy one each of the building's three techniques in the other three buildings. (Do this as close to simultaneously as you can. Use the "yinyang" cheat to get enough yin or yang points.)			
Second-level unit enters building?			
Second-level unit progress meter OK?			
Technique 1 progress meter OK?			
Technique 2 progress meter OK?			
Technique 3 progress meter OK?			
Third-level unit exits building?			

Now, send a peasant into the first building, a first-level unit into the second building, and a second-level unit into the third. (Do this as close to simultaneously as you can.)			
Peasant enters building?			
Progress meter OK?			
First-level unit enters building?			
Progress meter OK?			
Second-level unit enters building?			
Progress meter OK?			
New units exit buildings?			

Appendix C

Basic Test Plan Template

Game Name

1. Copyright Information

Table of Contents

SECTION I: QA TEAM (and areas of responsibility)

1. QA Lead
    a. Office phone
    b. Home phone
    c. Mobile phone
    d. Email / IM / Discord / Slack, etc. addresses
2. Internal Testers
3. External Test Resources

SECTION II: TESTING PROCEDURES

1. General Approach
   a. Basic Responsibilities of Test Team
      i. Bugs
         1. Detect them as soon as possible after they enter the build
         2. Research them
         3. Communicate them to the dev team
         4. Help get them resolved
         5. Track them
      ii. Maintain the Daily Build
      iii. Levels of Communication. There's no point in testing unless the results of the tests are communicated in some fashion. There are a range of possible outputs from QA. In increasing levels of formality, they are as follows:
         1. Conversation
         2. ICQ/IM/Chat/Discord/Slack
         3. Email to individual
         4. Email to group
         5. Daily Top Bugs list
         6. Stats/Info Dump area on Dev Site
         7. Formal entry into Bug Tracking System
2. Daily Activities
   a. The Build
      i. Generate a daily build.
      ii. Run the daily regression tests, as described in "Daily Tests" which follows.
      iii. If everything is OK, post the build so that everyone can get it.

iv. If there's a problem, send an email message to the entire dev team that the new build cannot be copied, and contact whichever developers can fix the problem.

v. Decide whether a new build needs to be run that day.

b. Daily Tests

i. Run through a predetermined set of single-player levels, performing a specified set of activities.

1. Level #1
   a. Activity #1
   b. Activity #2
   c. Additional activities, as needed
   d. The final activity is usually to run an automated script that reports the results of the various tests and posts them in the QA portion of the internal website.
2. Level #2
3. Additional levels, as needed

ii. Run through a predetermined set of multiplayer levels, performing a specified set of activities.

1. Level #1
   a. Activity #1
   b. Activity #2
   c. Additional activities, as needed
   d. The final activity is usually to run an automated script that reports the results of the various tests and posts them in the QA portion of the internal website.
2. Level #2
3. Additional levels, as needed

   iii. Email showstopper crashes or critical errors to the entire team.
   iv. Post showstopper crashes or critical errors to the daily top bugs list (if one is being maintained).

3. Daily Reports
   a. Automated reports from the preceding daily tests are posted in the QA portion of the internal website.
4. Weekly Activities
   a. Weekly Tests
      i. Run through every level in the game (not just the preset ones used in the daily test), performing a specified set of activities and generating a predetermined set of tracking statistics. The same machine should be used each week.
         1. Level #1
            a. Activity #1
            b. Activity #2
            c. Additional activities, as needed
         2. Level #2
         3. Additional levels, as needed
      ii. Weekly review of bugs in the Bug Tracking System
         1. Verify that bugs marked "fixed" by the development team really are fixed.
         2. Check the appropriateness of bug rankings relative to where the project is in the development.
         3. Acquire a "feel" for the current state of the game, which can be communicated in discussions to the producer and department heads.
         4. Generate a weekly report of closed bugs.
   b. Weekly Reports
      i. Tracking statistics, as generated in the weekly tests.

5. Ad Hoc Testing
    a. Perform specialized tests as requested by the producer, tech lead, or other development team members.
    b. Determine the appropriate level of communication to report the results of those tests.
6. Integration of Reports from External Test Groups
    a. If at all possible, ensure that all test groups are using the same bug tracking system.
    b. Determine which group is responsible for maintaining the master list.
    c. Determine how frequently to reconcile bug lists against each other.
    d. Ensure that only one consolidated set of bugs is reported to the development team.
7. Focus Testing (if applicable)
    a. Recruitment methods
    b. Testing location
    c. Who observes them?
    d. Who communicates with them?
    e. How is their feedback recorded?
8. Compatibility Testing (if applicable)
    a. Selection of external vendor
    b. Evaluation of results
    c. Method of integrating filtered results into bug tracking system
9. Test Automation (if applicable)

SECTION III: HOW TESTING REQUIREMENTS ARE GENERATED

1. Some requirements are generated by this plan.
2. Requirements can also be generated during project meetings, or other formal meetings held to review current priorities (such as the set of predetermined levels used in the daily tests).
3. Requirements can also result from changes in a bug's status within the bug tracking system. For example, when a bug is marked "fixed" by a developer, a requirement is generated for someone to verify that it has been truly fixed and can be closed out. Other status changes include "Need More Info" and "Can't Duplicate," each of which creates a requirement for QA to investigate the bug further.
    a. Some requirements are generated when a developer wants QA to check a certain portion of the game (see earlier "Ad Hoc Testing" section).

SECTION IV: BUG TRACKING SOFTWARE

1. Package name
2. How many seat licenses will be needed for the project?
3. Access instructions (everyone on the team should have access to the bug database)
4. "How to report a bug" instructions for using the system

SECTION V: BUG CLASSIFICATIONS

1. "A" bugs and their definition
2. "B" bugs and their definition
3. "C" bugs and their definition

SECTION VI: BUG TRACKING

1. Who classifies the bug?
2. Who assigns the bug?
3. What happens when the bug is fixed?
4. What happens when the fix is verified?

SECTION VII: SCHEDULING AND LOADING

1. Rotation Plan: How testers will be brought on and off the project, so that some testers stay on it throughout its life cycle while "fresh eyes" are periodically brought in
2. Loading Plan: Resource plan that shows how many testers will be needed at various points in the life of the project

SECTION VIII: EQUIPMENT BUDGET AND COSTS

1. QA Team Personnel with Hardware and Software Toolset
    a. Team Member #1
        i. Hardware
            1. Testing PC
                a. Specs
            2. Console debug kit
                a. Add-ons (TV, controllers, etc.)
            3. Record/capture hardware or software
        ii. Software Tools Needed
            1. Bug tracking software
            2. Other
    b. Team Member #2
    c. Additional Team Members, as needed
2. Equipment Acquisition Schedule and Costs (summary of who needs what, when they will need it, and how much it will cost)

Appendix D

COMBINATORIAL TEST TEMPLATES

Tables of Parameters with Two Test Values

TABLE D.1 Three parameters, two values each

Test	ParamA	ParamB	ParamC
1	A1	B1	C1
2	A2	B1	C2
3	A1	B2	C2
4	A2	B2	C1

TABLE D.2 Four parameters, two values each

Test	ParamA	ParamB	ParamC	ParamD
1	A1	B1	C1	D1
2	A2	B1	C2	D1
3	A1	B2	C2	D2
4	A2	B2	C1	D1
5	A2	B1	C1	D2

TABLE D.3 Five parameters, two values each

Test	ParamA	ParamB	ParamC	ParamD	ParamE
1	A1	B1	C1	D1	E1
2	A2	B1	C2	D1	E1
3	A1	B2	C2	D2	E2
4	A2	B2	C1	D1	E2
5	A2	B1	C1	D2	E2
6	A*	B2	C*	D2	E1

TABLE D.4 Six parameters, two values each

Test	ParamA	ParamB	ParamC	ParamD	ParamE	ParamF
1	A1	B1	C1	D1	E1	F1
2	A2	B1	C2	D1	E1	F1
3	A1	B2	C2	D2	E2	F1
4	A2	B2	C1	D1	E2	F2
5	A2	B1	C1	D2	E2	F2
6	A1	B2	C2	D2	E1	F2

TABLE D.5 Seven parameters, two values each

Test	ParamA	ParamB	ParamC	ParamD	ParamE	ParamF	ParamG
1	A1	B1	C1	D1	E1	F1	G1
2	A2	B1	C2	D1	E1	F1	G2
3	A1	B2	C2	D2	E2	F1	G2
4	A2	B2	C1	D1	E2	F2	G2
5	A2	B1	C1	D2	E2	F2	G1
6	A1	B2	C2	D2	E1	F2	G1

TABLE D.6 Eight parameters, two values each

Test	Param A	Param B	Param C	Param D	Param E	Param F	Param G	Param H
1	A1	B1	C1	D1	E1	F1	G1	H1
2	A2	B1	C2	D1	E1	F1	G2	H2
3	A1	B2	C2	D2	E2	F1	G2	H1
4	A2	B2	C1	D1	E2	F2	G2	H2
5	A2	B1	C1	D2	E2	F2	G1	H1
6	A1	B2	C2	D2	E1	F2	G1	H2

TABLE D.7 Nine parameters, two values each

Test	Param A	Param B	Param C	Param D	Param E	Param F	Param G	Param H	Param J
1	A1	B1	C1	D1	E1	F1	G1	H1	J1
2	A2	B1	C2	D1	E1	F1	G2	H2	J2
3	A1	B2	C2	D2	E2	F1	G2	H1	J2
4	A2	B2	C1	D1	E2	F2	G2	H2	J1
5	A2	B1	C1	D2	E2	F2	G1	H1	J2
6	A1	B2	C2	D2	E1	F2	G1	H2	J1

TABLE D.8 Ten parameters, two values each

Test	Param A	Param B	Param C	Param D	Param E	Param F	Param G	Param H	Param J	Param K
1	A1	B1	C1	D1	E1	F1	G1	H1	J1	K1
2	A2	B1	C2	D1	E1	F1	G2	H2	J2	K2
3	A1	B2	C2	D2	E2	F1	G2	H1	J2	K1
4	A2	B2	C1	D1	E2	F2	G2	H2	J1	K1
5	A2	B1	C1	D2	E2	F2	G1	H1	J2	K2
6	A1	B2	C2	D2	E1	F2	G1	H2	J1	K2

Tables of Parameters with Three Test Values

TABLE D.9 Three parameters, three values each

Test	ParamA	ParamB	ParamC
1	A1	B1	C1
2	A2	B2	C2
3	A3	B3	C3
4	A1	B2	C3
5	A2	B3	C1
6	A3	B1	C2
7	A1	B3	C2
8	A2	B1	C3
9	A3	B2	C1

TABLE D.10 Two parameters with three values, one parameter with two values

Test	ParamA	ParamB	ParamC
1	A1	B1	C1
2	A2	B2	C1
3	A3	B3	C1
4	A1	B2	C2
5	A2	B3	C2
6	A3	B1	C2
7	A1	B3	C*
8	A2	B1	C*
9	A3	B2	C*

TABLE D.11 One parameter with three values, two parameters with two values

Test	ParamA	ParamB	ParamC
1	A1	B1	C1
2	A2	B2	C1
3	A3	B1	C1
4	A1	B2	C2
5	A2	B1	C2
6	A3	B2	C2

TABLE D.12 Four parameters, three values each

Test	ParamA	ParamB	ParamC	ParamD
1	A1	B1	C1	D1
2	A2	B2	C2	D1
3	A3	B3	C3	D1
4	A1	B2	C3	D2
5	A2	B3	C1	D2
6	A3	B1	C2	D2
7	A1	B3	C2	D3
8	A2	B1	C3	D3
9	A3	B2	C1	D3

TABLE D.13 Three parameters with three values, one parameter with two values

Test	ParamA	ParamB	ParamC	ParamD
1	A1	B1	C1	D1
2	A2	B2	C2	D1
3	A3	B3	C3	D1
4	A1	B2	C3	D2
5	A2	B3	C1	D2
6	A3	B1	C2	D2
7	A1	B3	C2	D*
8	A2	B1	C3	D*
9	A3	B2	C1	D*

TABLE D.14 Two parameters with three values, two parameters with two values

Test	ParamA	ParamB	ParamC	ParamD
1	A1	B1	C1	D1
2	A2	B2	C2	D1
3	A3	B3	C1	D1
4	A1	B2	C1	D2
5	A2	B3	C2	D2
6	A3	B1	C2	D2
7	A1	B3	C2	D*
8	A2	B1	C1	D*
9	A3	B2	C*	D*

TABLE D.15 One parameter with three values, three parameters with two values

Test	ParamA	ParamB	ParamC	ParamD
1	A1	B1	C1	D1
2	A2	B2	C2	D1
3	A3	B1	C2	D2
4	A1	B2	C2	D2
5	A2	B1	C1	D2
6	A3	B1	C1	D1

TABLE D.16 Three parameters with three values, two parameters with two values

Test	ParamA	ParamB	ParamC	ParamD	ParamE
1	A1	B1	C1	D1	E1
2	A2	B2	C2	D2	E2
3	A3	B3	C3	D1	E2
4	A1	B2	C3	D2	E1
5	A2	B3	C1	D2	E1
6	A3	B1	C2	D2	E1
7	A1	B3	C2	D1	E2
8	A2	B1	C3	D1	E2
9	A3	B2	C1	D1	E2

TABLE D.17 Two parameters with three values, three parameters with two values

Test	ParamA	ParamB	ParamC	ParamD	ParamE
1	A1	B1	C1	D1	E1
2	A2	B2	C2	D2	E1
3	A3	B3	C1	D2	E2
4	A1	B2	C2	D1	E2
5	A2	B3	C2	D1	E2
6	A3	B1	C2	D1	E2
7	A1	B3	C1	D2	E1
8	A2	B1	C1	D2	E1
9	A3	B2	C1	D2	E1

TABLE D.18 One parameter with three values, four parameters with two values

Test	ParamA	ParamB	ParamC	ParamD	ParamE
1	A1	B1	C1	D1	E1
2	A2	B2	C2	D1	E1
3	A3	B1	C2	D2	E1
4	A1	B2	C2	D2	E2
5	A2	B1	C1	D2	E2
6	A3	B2	C1	D1	E2

TABLE D.19 Three parameters with three values, three parameters with two values

Test	ParamA	ParamB	ParamC	ParamD	ParamE	ParamF
1	A1	B1	C1	D1	E1	F1
2	A2	B2	C2	D2	E2	F1
3	A3	B3	C3	D1	E2	F2
4	A1	B2	C3	D2	E1	F2
5	A2	B3	C1	D2	E1	F2
6	A3	B1	C2	D2	E1	F2
7	A1	B3	C2	D1	E2	F1
8	A2	B1	C3	D1	E2	F1
9	A3	B2	C1	D1	E2	F1

TABLE D.20 Three parameters with three values, four parameters with two values

Test	ParamA	ParamB	ParamC	ParamD	ParamE	ParamF	ParamG
1	A1	B1	C1	D1	E1	F1	G1
2	A2	B2	C2	D2	E2	F1	G1
3	A3	B3	C3	D1	E2	F2	G1
4	A1	B2	C3	D2	E1	F2	G2
5	A2	B3	C1	D2	E1	F2	G*
6	A3	B1	C2	D2	E1	F2	G*
7	A1	B3	C2	D1	E2	F1	G2
8	A2	B1	C3	D1	E2	F1	G2
9	A3	B2	C1	D1	E2	F1	G2

Appendix E

Test Flow Diagram (TFD) Templates

Power-Ups

Power-ups are items that give your character some kind of temporary bonus. You might need to drive over them, run over them, trigger a special item in a puzzle, or hit a special sequence on your game controller or keypad. The TFD template in Figure E.1 covers acquiring the power-up, using its abilities, canceling the power-up, checking for power-up expiration, and stacking power-ups. This same template could also be used for RPG and adventure games, where a player can trigger temporary effects from a weapon,

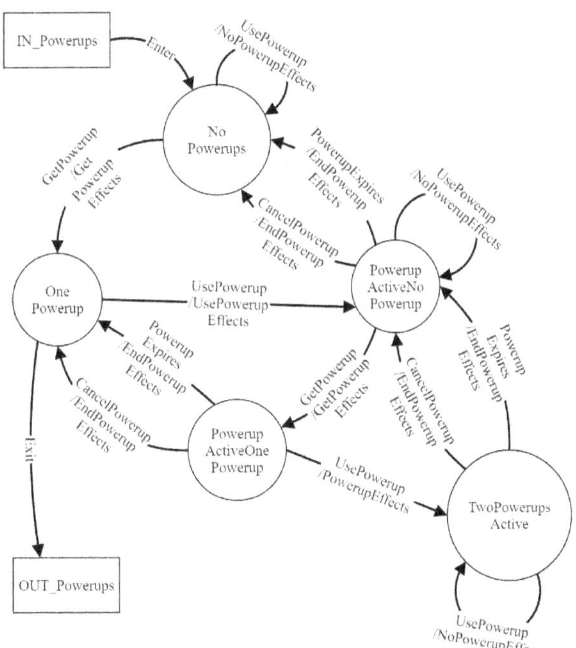

FIGURE E.1 Power-ups TFD template

get a temporary boost from an item, or receive temporary "buff" spells from other characters.

Craft Item

Crafting an item in a game world requires the player to have the ingredients and the skill to craft that particular type of item. Besides being trained in the right skill, the character must also have raised his skill to a sufficient level to make a crafting attempt of the target item. Some or all of the ingredients are normally consumed, whether or not the crafting attempt was successful. These factors are incorporated into the TFD template in Figure E.2.

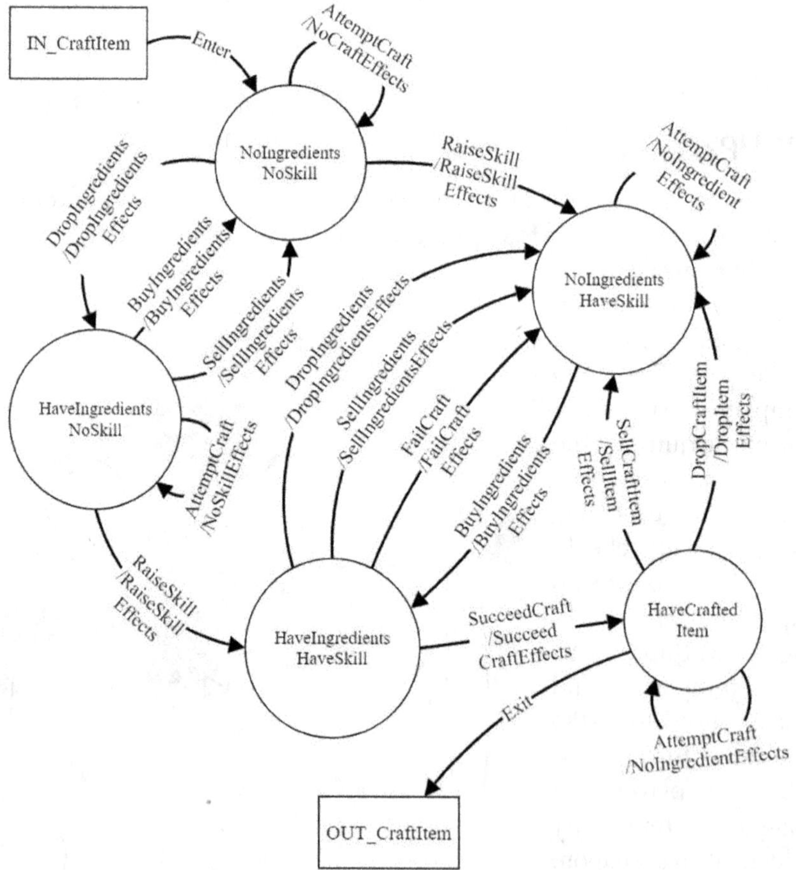

FIGURE E.2 Craft Item TFD template

Heal Character

Whether it is medics, magic, or a well-deserved nap, nothing beats a timely heal to get you through a tough mission, level, or battle. Get a friend to resurrect you or respawn to start over. You can also change "Heal" to "Repair," and use the TFD template in Figure E.3 when it is your car or robot that is taking a beating.

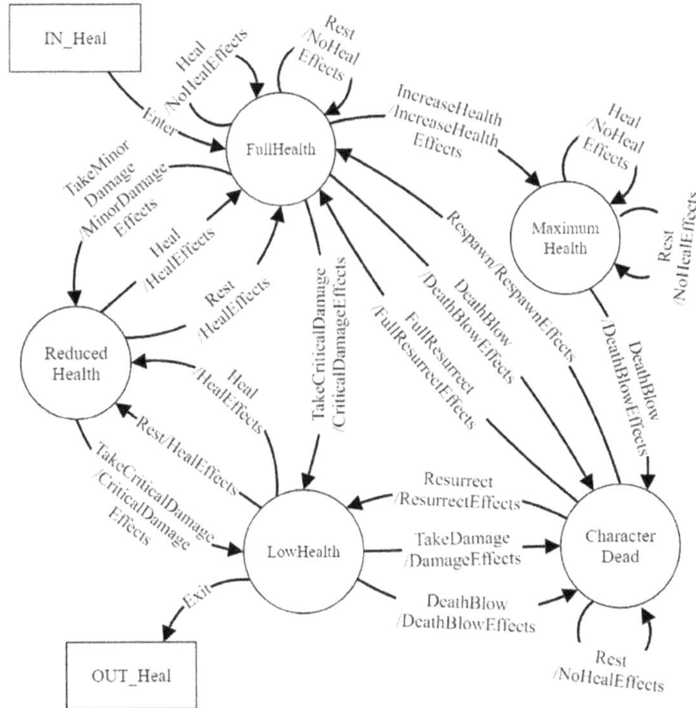

FIGURE E.3 Heal Character TFD template

Create/Save

Games are full of custom elements. You can create characters, teams, playbooks, song lists, and skateboards. You also need to save them if you want to see them the next time you fire up the game. The TFD template in Figure E.4 handles creating, deleting, filling up your save slots, and restarting the game without saving your changes. If you are using this for something besides character creation, replace "Character" with the name of the type of element you are testing.

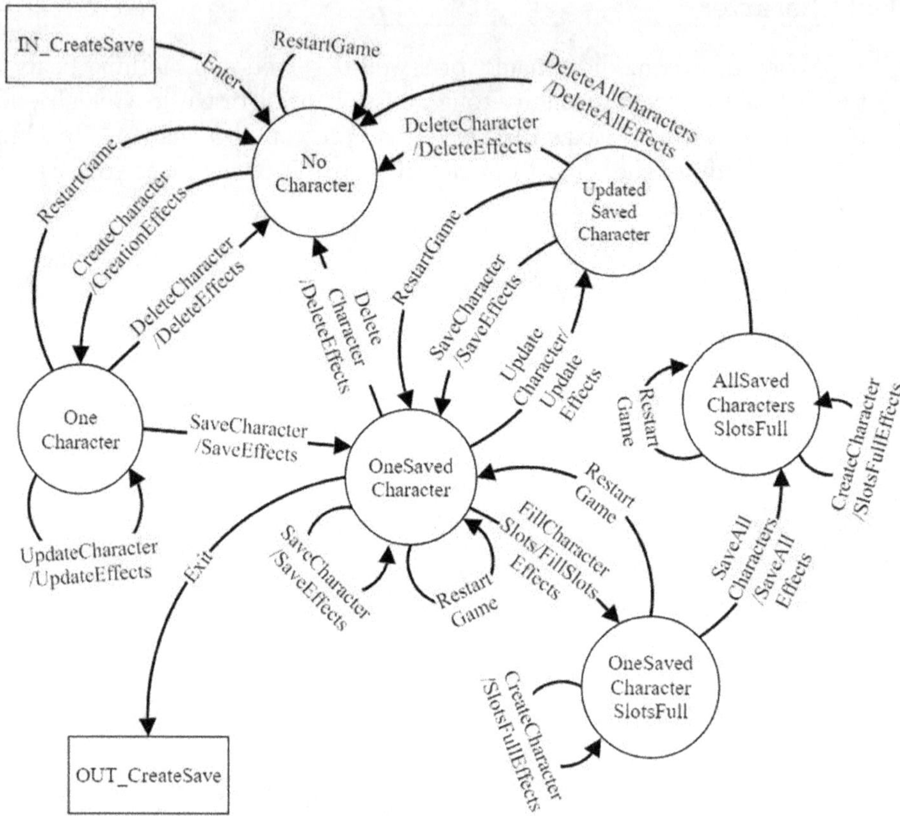

FIGURE E.4 Create/Save TFD template

Unlock and Buy Item

Simulation, RPG, adventure, and even sports games tend to have featured items that you can purchase once you have unlocked the ability to purchase the item and have enough points to actually buy it. The "items" could be weapons, spells, clothing, furniture, mini-games, vehicles, or new levels. To unlock them, you might have to complete a specific task or mission, defeat a particular opponent, raise your character's level, or achieve a result under special circumstances. Test your purchasing power using the TFD template in Figure E.5.

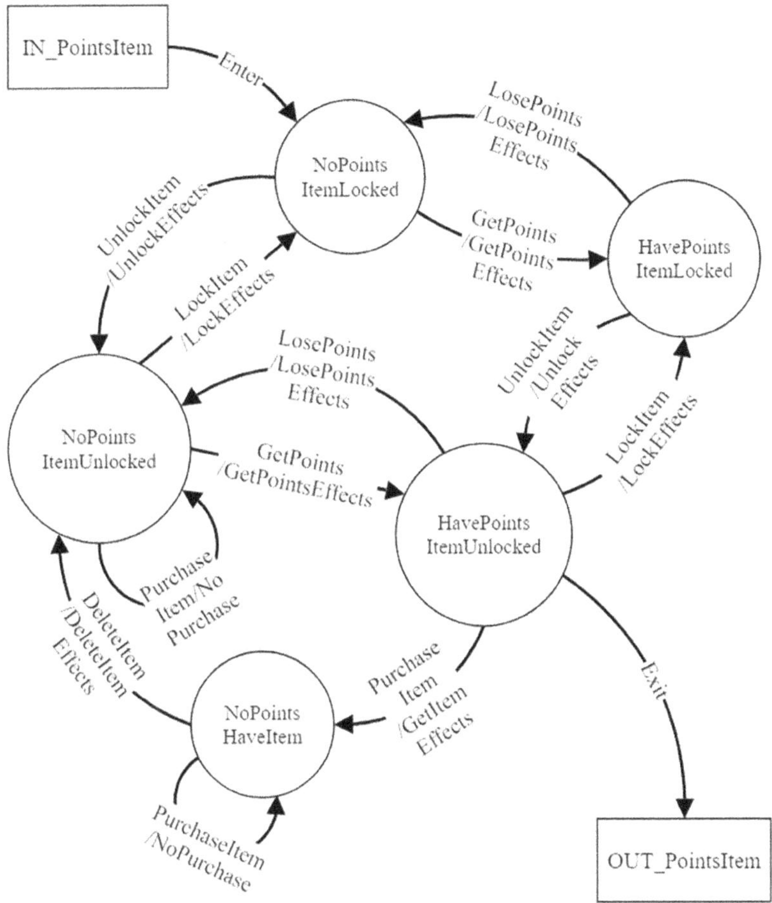

FIGURE E.5 Use Points to Buy Unlocked Item TFD template

Update Song List

It is very effective when games incorporate popular music. You might find today's hits blasting from a car radio or a street basketball court. Music can also be a more integral part of gameplay, such as in dancing, musical instrument, or karaoke games. The TFD template in Figure E.6 reflects the player's ability to add and delete songs, order them, map them to game events, and trigger them from within the game. Depending on the game, triggering could be user-controlled, such as tuning to a particular in-game radio station, or event-driven, such as the music played when the home team scores a touchdown.

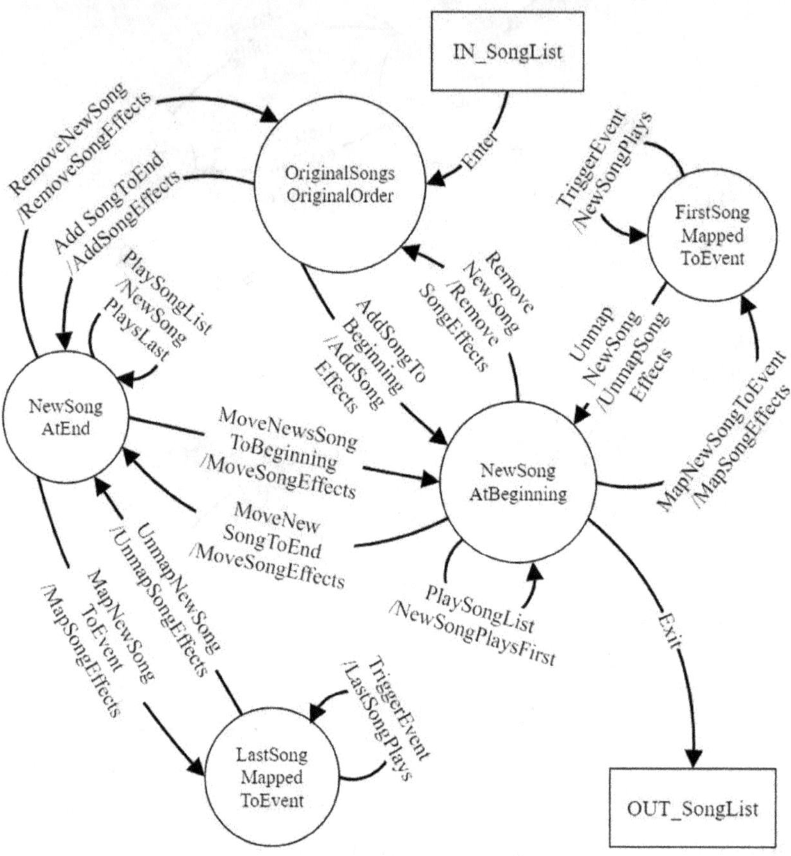

FIGURE E.6 Update Song List TFD template

Complete a Mission or Quest

Many games will reward points, money, items, or access to new parts of the game if you can complete a particular mission, quest, or other designated goal. It is common for these missions to be broken into multiple objectives that must be completed individually to succeed and earn the reward. These objectives could be things such as capturing a set of territories or villains, winning a series of competitions, or completing a set of bonus words. The TFD template in Figure E.7 is constructed for goals with three objectives, but you can also use it for two objectives by knocking out the states and flows that deal with "Objective3."

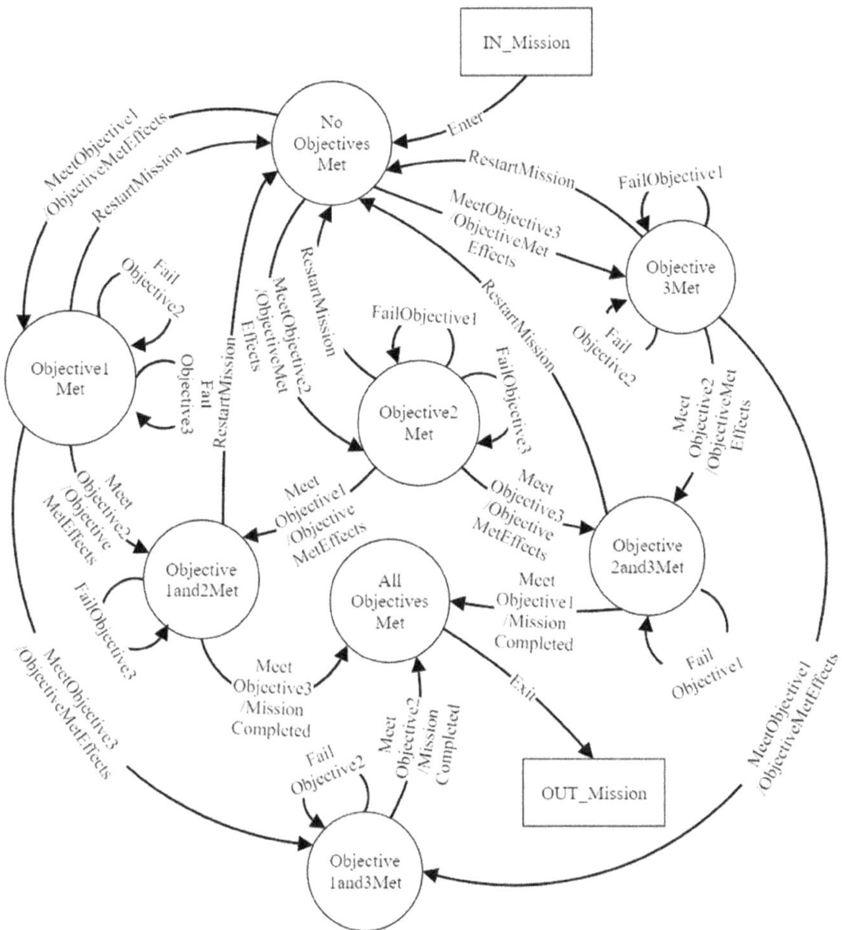

FIGURE E.7 Complete a Mission or Quest TFD template

Get Weapon and Ammo

The TFD template in Figure E.8 is an enhancement of the diagram from the walkthrough in Chapter 9. A state and flows have been added for handling the case where the weapon has maximum ammo. You can also apply this TFD structure to game elements which have a similar relationship, such as cars and fuel or spells and mana. Just replace "Gun" and "Ammo" with the corresponding elements.

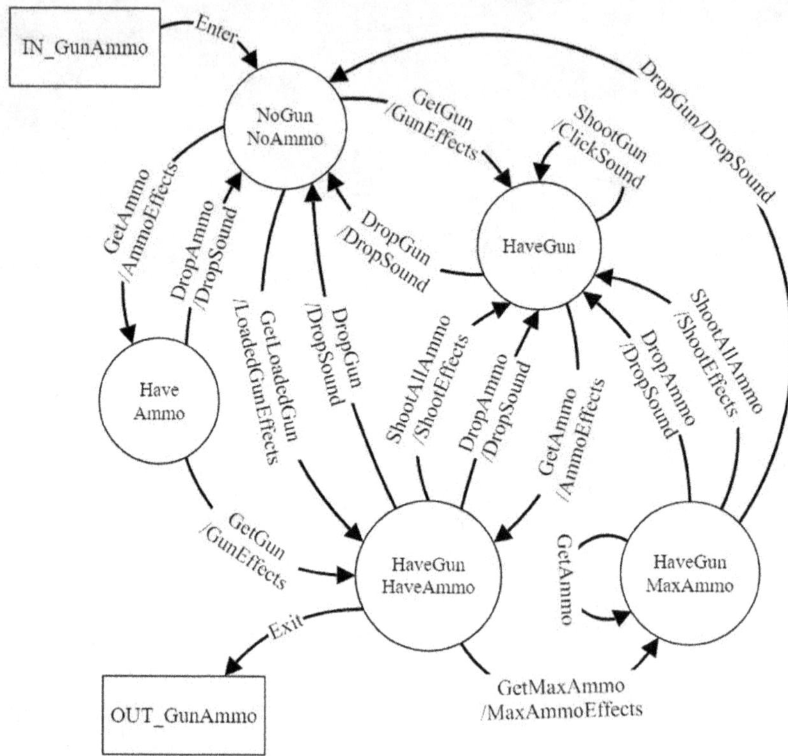

FIGURE E.8 Weapon and Ammo TFD template

Glossary

AI Artificial intelligence. In game development, this abbreviation for artificial intelligence most commonly refers to a simple scripted behavior within a game, or a bot or NPC whose behavior is controlled by a script or algorithm.

AAA Also called "Triple A." These are the types of ultra-high-budget games that take years to produce and require the coordination of hundreds of developers and testers across multiple studios worldwide. Major releases in the *Call of Duty*, *Final Fantasy*, or *God of War* franchises are examples of "AAA" games.

Alpha The first major milestone in the development of a video game. Typically, this is a version that has code for all features implemented and working.

App A self-installed, single-purpose software application, or program, especially on mobile or tablet devices. Games can be apps. Not all apps are games.

Asset Anything in a video game that you can see, hear, or read. Assets are datasets that the game code calls upon to create the content of the game. Assets can include textures, animations, audio files, cut scenes, backgrounds, environments, sprites, and dialog text.

Beta	The second major milestone in the development of a video game. Typically, this a version that has all content implemented and playable.
Bot	A character or unit controlled by the game program, usually via a simple script or algorithm.
Box	A computer or game console.
Build	A distinct work-in-progress version of a program or video game. Each successive build can contain more code, more features, or assets, and fixes to bugs found in prior builds.
Code	Instructions to a computer written in a programming language. Code and assets are compiled (combined) into an executable video game or program.
Compatibility Testing	Testing to see that the game is playable on different operating systems and browsers.
Configuration Testing	Testing the game on a particular piece of computer or console hardware, or any combination of hardware and peripheral devices.
Console	A game console is a mass-produced purpose-built computer optimized for playing video games on a television or computer monitor. Within a video game or computer program, the console is an interface in which the user can input specific text commands, such as the Command Prompt in Windows operating systems.
CPU	Although this refers to the Central Processing Unit of a computer, it is commonly used as a synonym for AI or game scripting. Players can sometimes choose a "CPU opponent" in multiplayer games.
Developer	A business whose primary focus is making video games. A developer can also be a programmer, artist, or any other member of the team developing a game.
Exploit	The use of a bug by a player to gain an advantage in a video game. Exploits can include a player using a hole in the level geometry to shoot other players from hiding, or using a means of duplicating valuable items for free.

First Playable A prototype version of a video game. It is often the first version that can be played. Earlier versions might include non-interactive proof-of-technology demos.

GDD Game Design Document. This document (or wiki) describes in detail the complete game experience, including gameplay systems, narrative context, art direction, and sound design. Bugs arise when the game behaves in ways that depart from the design articulated in the GDD.

Gold Master (Candidate) The third and final major milestone in the development of a video game.

GUI Graphical User Interface.

Interface An interface occurs at any point where information or commands are transferred or exchanged between the computer and a player or user. Although a game controller is a type of interface, this term is commonly used when referring to on-screen interactions, as in “graphical user interface” or GUI.

Live Version The current version of an online video game that is released and playable by the general public.

Localization This is the process of preparing a video game created in one country for release in other countries or regions. Rather than simply “translating,” localization takes into account such concerns as cultural taboos and business regulations in the target market.

Machine A computer or game console.

MMORPG Massively-Multiplayer Online Role-Playing Game. These are live-service games that connect hundreds of players at once in a shared environment.

NPC Non-Player Character, usually controlled by a simple script or algorithm.

Primary Tester The assistant to the test lead, who helps to manage the test team.

Publisher A business whose primary focus is the financing and marketing of video games. Publishers may own their own "internal" development teams to make games, or they may hire third-party "external" studios to make games.

Smoke Testing This is a test (or tests) of a new build to determine whether it can be installed, launched, and is stable enough to distribute to the full test team.

SQA Software Quality Assurance includes all the methods of assuring that a program is being developed using processes that conform to high standards or goals. "White Box" testing is a common SQA technique.

SQAP Software Quality Assurance Plan. This is a documented plan for how software quality will be monitored and tracked during the project. This is different from the QA Plan or Test Plan, which documents how the game will be tested.

Street Date The date that a video game is available for purchase by the general public.

Studio A game development company, or a game development team owned by a publisher.

TDD Technical Design Document. This document (or wiki) describes in detail the technical considerations, such as in-game physics systems and graphics programming solutions, that the engineering team will be using to create the game on the target platform(s). Bugs arise when the software behaves in ways that depart from the design articulated in the TDD.

Test Lead The person in charge of a test team. Can also be known as "QA Lead" or "Lead Tester".

Unit Testing Testing code from a small part of a game or program before it is integrated into the full project.

INDEX

R

S

T

U

V

W

www.ingramcontent.com/pod-product-compliance
Lightning Source LLC
Chambersburg PA
CBHW080138220326
41598CB00032B/5107

* 9 7 8 1 5 0 1 5 2 1 6 8 3 *